WASHINGTON ROCKS!

A GUIDE TO GEOLOGIC SITES IN THE EVERGREEN STATE

WASHINGTON ROCKS!

A GUIDE TO GEOLOGIC SITES IN THE EVERGREEN STATE

Eugene Kiver, Chad Pritchard, and Richard Orndorff

2016
Mountain Press Publishing Company
Missoula, Montana

Photos by authors unless otherwise credited.
Cover photo by Barbara Kiver.

GEOLOGY ROCKS!

A state-by-state series that introduces readers to some of the
most compelling and accessible geologic sites in each state.

 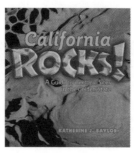 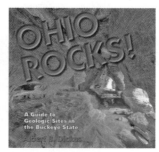

Library of Congress Cataloging-in-Publication Data

Names: Kiver, Eugene P. | Pritchard, Chad, 1979- | Orndorff, Richard L.
Title: Washington rocks! : a guide to geologic sites in the Evergreen State /
  Eugene Kiver, Chad Pritchard, and Richard Orndorff.
Description: Missoula, Montana : Mountain Press Publishing Company, 2016.
Identifiers: LCCN 2016000386 | ISBN 9780878426546 (pbk. : alk. paper)
Subjects: LCSH: Geology—Washington (State)—Guidebooks. | Washington
  (State)—Guidebooks.
Classification: LCC QE175 .K58 2016 | DDC 557.97—dc23
LC record available at http://lccn.loc.gov/2016000386

PRINTED IN THE UNITED STATES

MP **Mountain Press**
PUBLISHING COMPANY
P.O. Box 2399 • Missoula, MT 59806 • 406-728-1900
800-234-5308 • info@mtnpress.com
www.mountain-press.com

View to the north of the northern end of Steamboat Rock in the early morning light. Massive layers of Columbia River Basalt overlie granite, the light-colored, rounded rock visible at the base of the lower talus slopes.

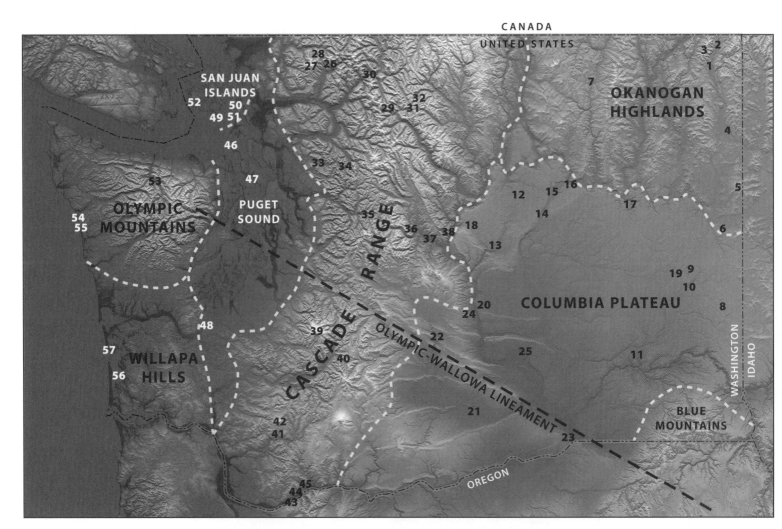

The sites in the book are organized into seven chapters according to geologic and geomorphic provinces: Okanogan Highlands, Columbia Plateau, Cascade Range, Puget Sound, San Juan Islands, Olympic Mountains, and Willapa Hills.

CONTENTS

PREFACE

Each of the fifty-seven sites described in this book is a silent monument to geologic events of the past and sometimes of the present and future. The geologic summaries of the sites are written for the nongeologist and are necessarily short, intended to stimulate minds and encourage people to visit these special places to examine firsthand the geologic history of Washington. The fifty-seven sites are accessible to the public, most in national parks, state parks, national forests, recreation areas, and other public parks and land. Special care should be exercised when visiting sites along roadways. Providing complete details of all of the remarkable events that combine to establish the diverse geologic features of Washington is beyond the scope of this book; see the References and Further Reading section for many good reads about Washington geology.

We wish to thank our colleagues, friends, and acquaintances who provided us with a number of outstanding diagrams, photos, and information that assisted us in understanding the complex geology of Washington State. Brian Atwater graciously shared some of his insights into the great earthquake history along the Pacific Coast, and Pat Pringle spent considerable time tweaking his illustration of a subduction zone along the Washington coast to include in the book. Judy McMillan and Larry Conboy assisted with photo preparation and some of the diagrams. Scott Peterson reviewed much of the introductory chapter.

Paul Weis's and Dale Stradling's photo collections were made available to us prior to their deaths. Others graciously provided important photographs to help better tell the story of the long geologic history of the state and the geomorphic processes that have shaped our landscapes. Photos by Bruce Bjornstad, John Scurlock, Ned Brown, Bill Baccus, Meghan Lunney, and Cascade Cliffs Vineyard and Winery were particularly helpful. John Whitmer and Dave Tucker were helpful in gathering information and photographs.

Special thanks to Barbara Kiver for many excellent photographs and to Mark Kiver Photography for use of some outstanding images. Understanding and encouragement from our wives was particularly important in the successful completion of this book.

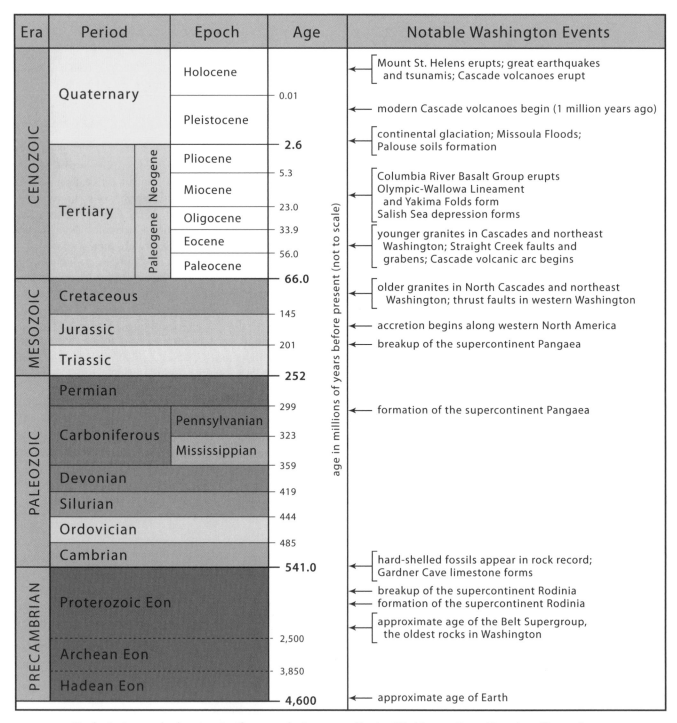

Era	Period	Epoch	Age	Notable Washington Events
CENOZOIC	Quaternary	Holocene		Mount St. Helens erupts; great earthquakes and tsunamis; Cascade volcanoes erupt
			0.01	modern Cascade volcanoes begin (1 million years ago)
		Pleistocene		continental glaciation; Missoula Floods; Palouse soils formation
			2.6	
	Tertiary (Neogene)	Pliocene		Columbia River Basalt Group erupts Olympic-Wallowa Lineament and Yakima Folds form Salish Sea depression forms
			5.3	
		Miocene		
			23.0	
	Tertiary (Paleogene)	Oligocene		younger granites in Cascades and northeast Washington; Straight Creek faults and grabens; Cascade volcanic arc begins
			33.9	
		Eocene		
			56.0	
		Paleocene		
			66.0	
MESOZOIC	Cretaceous			older granites in North Cascades and northeast Washington; thrust faults in western Washington
			145	accretion begins along western North America
	Jurassic		201	breakup of the supercontinent Pangaea
	Triassic			
			252	
PALEOZOIC	Permian		299	formation of the supercontinent Pangaea
	Carboniferous	Pennsylvanian	323	
		Mississippian	359	
	Devonian		419	
	Silurian		444	
	Ordovician		485	
	Cambrian		541.0	hard-shelled fossils appear in rock record; Gardner Cave limestone forms
PRECAMBRIAN	Proterozoic Eon			breakup of the supercontinent Rodinia
				formation of the supercontinent Rodinia
				approximate age of the Belt Supergroup, the oldest rocks in Washington
			2,500	
	Archean Eon		3,850	
	Hadean Eon		4,600	approximate age of Earth

age in millions of years before present (not to scale)

Geologic time scale showing significant geologic events affecting Washington State. Time, in millions of years before present, is from the 2012 Geological Society of America time scale.

BRIEF HISTORY OF WASHINGTON'S GEOLOGY

Washington has been at the edge of a tectonic plate throughout its entire geologic history. Nearly 1 billion years ago, seawater lapped against a coastline that existed where Spokane now sits in eastern Washington. The rest of Washington has been added to the edge of the North American continent, bit by bit, as pieces of land and ocean floor smashed into the coast from the west and volcanoes erupted lava and ash. Oceanic sediments recently added to the modern coast have been crumpled into the still-rising Olympic Mountains, a sign that Washington continues to add real estate along its coast.

Subduction zones along the western edge of North America have been responsible for much of the growth. The surface of Earth is made of seven major plates and many minor ones, some continental and others oceanic. As the rigid plates move about the Earth, they interact with each other at boundaries. Where oceanic plates collide with continental plates, the denser oceanic plate is subducted beneath the continental plate along subduction zones. Islands, microcontinents, and ocean sediments associated with the oceanic plate are not dense enough to get subducted, so

they often get added to the edge of the continent. Today, what's left of the Juan de Fuca Plate is subducting beneath the North American Plate along the Cascadia subduction zone.

Subduction leads to partial melting of the Earth's upper mantle, and the magma rises and erupts as volcanoes. Magma generated by the subduction of the Juan de Fuca Plate produces the active volcanoes of the modern Cascade Range. Some of the largest known earthquakes also occur along subduction zones. The most recent earthquake of magnitude 8 or higher in the Pacific Northwest was on January 26, 1700.

While the continent grew along its edge, volcanic arcs came and went, magma intruded, huge volumes of lava occasionally erupted, and ice sheets ebbed and flowed. The

Generalized plate tectonics diagram showing a spreading ridge and subduction zone and their relation to the modern Cascade Range volcanoes.
—Diagram by Pat Pringle

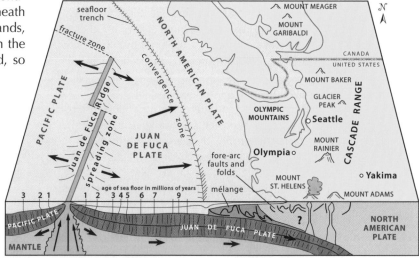

Plumes of buoyant asthenosphere undergo decompression melting to form new ocean crust (mainly basalt).

Partial melting of rock above subducting oceanic plate forms magma that fuels Cascade Range volcanoes (mostly andesitic).

1

story of Washington's geologic history could fill several volumes. What follows is a brief history through time of the geologic highlights of Washington State, so you have a sense of where each of the fifty-seven sites included in this book fits into the state's geologic history.

Precambrian through Paleozoic Time

Planet Earth formed about 4.6 billion years ago. Rocks older than 2 billion years are rare, though, because old rocks are constantly being eroded away or buried by thousands of feet of sediment or volcanic rock. The heat and pressure of being buried deep within Earth changes the rocks by metamorphism, or even melts them. The oldest rocks in Washington are part of the homegrown Belt Supergroup, which is about 1.6 to 1.4 billion years old. These slightly metamorphosed sedimentary rocks were deposited in shallow water on a coastal plain or continental shelf or perhaps in a large basin filled with lakes. About 1 billion years ago, the Belt rocks were part of the supercontinent Rodinia, one of perhaps four supercontinents that formed during the history of Earth. Rodinia split apart about 800 million years ago, dividing the Belt rocks in two. The Belt rocks in eastern Washington remained attached to the continental nucleus of North America, which at that time was moving eastward. The Belt rocks west of the split became part of another continent. Their present whereabouts are unknown but may be in northeast Russia, easternmost Asia, or even Australia!

Following the split of Rodinia, ocean waters washed against the new edge of the continent just west of the present location of Spokane, and sediments accumulated. The rocks that formed from these sediments are called the Windermere Group and are found from Colville in the northeastern corner of the state north into British Columbia. Later, Paleozoic sediments, some with abundant fossils, were deposited on the newly formed coastal plain. A shallow sea at the western edge of North America prevailed during the Paleozoic Era, from 541 to 252 million years ago, as the continental plate continued its eastward movement for hundreds of millions of years.

By about 300 million years ago the continents had reassembled into another supercontinent, called Pangaea, but it

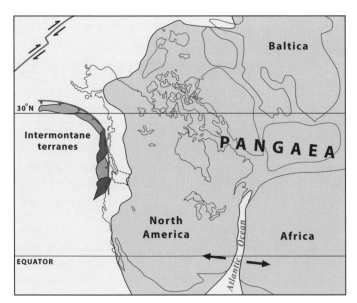

Pangaea began to split apart approximately 200 million years ago. The Intermontane terranes, which would accrete to North America in Mesozoic time, were located off the coast of the continent, and the coastline was approximately at the border of eastern Washington and Idaho. —Modified from Connor, 2014

too became unstable and split apart about 200 million years ago, sending various smaller plates outward to form the present continents of the world. The rift basin that formed where it split became the Atlantic Ocean. The North American Plate then moved westward, away from Europe and Africa, reversing its motion of the previous several hundred million years. That westward movement, which continues today, built the bewildering mass of mountain topography that characterizes the western side of North America.

Mesozoic Time

A subduction zone developed along the western edge of the westward-moving North American Plate. The subducting slab at that time is commonly referred to as the Farallon Plate, which dwarfed its current remnants, the Juan de Fuca,

Gorda, and Explorer Plates. Large blocks of island-like masses, or microcontinents, some sliced off distant continents by faults, were carried slowly north and eastward on the Farallon Plate toward the subduction zone, smashing into the western edge of the North American Plate. The addition of these exotic terranes, which were composed of volcanic island arcs, granitic rock, and sedimentary material, likely started around 180 to 160 million years ago and continued for a little over 100 million years. The slow but steady force of accreting terranes deformed rocks as far east as the Rocky Mountain states, thickened the crust, and generated large amounts of magma, forming the Idaho and Kaniksu batholiths. During approximately the same time, subduction zones generated the magma of the Sierra Nevada batholith of eastern California and the Coast Range batholith of British Columbia.

The Intermontane terranes started to accrete about 180 million years ago, when dinosaurs were still roaming the Earth. The oceanic crust between the continent and the exotic terranes crumpled during accretion and was uplifted in a long, narrow band between the early continent and the Intermontane terranes. This band of rock is called the Kootenay Arc and stretches from Washington north through western Canada. Although it is called an arc, it is not associated with a volcanic arc.

As the Intermontane terranes docked onto North America, a new subduction zone formed outboard of the Intermontane rocks near the present-day Okanogan Valley. The sediments shed off the subducting plate and overriding continent filled the trench, forming a crumpled wedge of sedimentary rock. We know the coastline at that time lay near the Methow Valley because rivers flowing into a sea deposited sediments there as recently as 120 to 105 million years ago.

The North Cascade microcontinent, a series of volcanic island chains riding on the Farallon Plate, was the next large package of land to arrive. The collision of the North Cascade microcontinent with the west edge of the Intermontane terranes began about 115 million years ago and produced

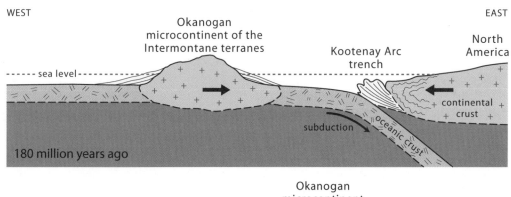

Cross sections showing accretion process of the Intermontane terranes welding onto North America about 160 million years ago. As the accretion progressed, the subduction zone shifted to the west to the vicinity of the Okanogan Valley. —Modified from Alt and Hyndman, 1984

3

magmatic intrusions and mountain building. The suture between the North Cascade microcontinent and rocks deposited on the continental shelf of the Intermontane terranes is the Ross Lake fault. East of the fault is the Methow terrane—the ocean floor sediments that were caught between the colliding landmasses in the Okanogan trench.

Farther west, the exotic rock packages in the western North Cascades and in the adjacent San Juan Islands were complexly folded, faulted, and metamorphosed as they collided with North America, until they resembled leaning slices of bread piled up against one another. Each slice represents a package of rock that slid beneath or on top of its neighbor along a thrust fault. Adjacent blocks have little resemblance to one another in type or age of rock, helping to account for the long time it took geologists to begin to unravel the jumbled geology of the area.

The slow yet steady collision of land generated heat and pressure that melted and metamorphosed rock. Starting about 90 million years ago large pools of magma well below the Earth's surface cooled into bodies of rock called plutons, which are composed of granite, granodiorite, and other igneous rocks. The age of the plutons and whether or not they are offset by some of the large faults in the western North Cascades have helped establish the chronology of geologic events.

Cenozoic Time

About 55 million years ago remnants of the Farallon Plate, the oceanic plate that had been subducting eastward under North America, began to move northeastward. The edge of North America was dragged to the north, forming large strike-slip faults

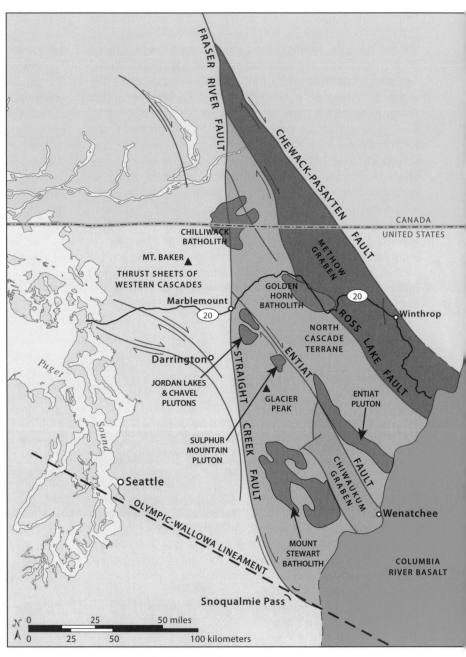

Major faults and grabens in the Cascade Range.

like the Straight Creek fault. In strike-slip faults, landmasses slide past one another mostly in a horizontal direction. The Straight Creek fault and its Canadian extension, the Fraser River fault, moved the intensely deformed western Cascade block northward about 70 miles, juxtaposing it against the gneisses and granites of the eastern part of the North Cascades. The motion stretched the crust to the east of the Ross Lake fault, and large down-dropped blocks, called grabens, formed between extensional faults. The Methow Valley in the eastern North Cascades is a graben, as is the Chiwaukum graben between Leavenworth and Wenatchee. The largest historic earthquake in Washington, estimated to be 7.0 magnitude or more, occurred in 1872 along the Spencer Canyon fault south of the town of Entiat within the Chiwaukum graben.

A prominent and enigmatic linear zone of weakness in the basement rock extends diagonally across Washington from the Olympic Mountains through Snoqualmie Pass to the Wallowa Mountains of northeastern Oregon. This topographic and structural feature, known as the Olympic-Wallowa Lineament and often referred to as the OWL, has also been called the Wallula fault zone. Specifically when and why the OWL formed is not known, but it was active during Cenozoic time, and even quite recently. The Yakima Folds, which formed in the 15-million-year-old Columbia River Basalt, have been bent by the OWL. The Straight Creek fault, which terminates against the OWL, was also bent by movement along this major structure.

Another major episode of melting and volcanism in Washington occurred about 50 million years ago, producing a younger set of granite plutons and eruptive centers across northeastern Washington and elsewhere in the western United States. This stage of volcanism was accompanied by crustal stretching. Rising masses of molten rock pushed upward, lifting once-deep crust and exposing highly metamorphosed rock in core complexes, such as the Okanogan, Kettle, and Spokane Domes. Rivers draining the volcanic mountains in eastern Washington deposited loads of sediment along the coast, which at that time was near Seattle. The Cascade Range did not exist yet, so there was no coastal barrier to the rivers.

Cascade Volcanism

The active volcanoes of the Cascade Range extend from southern British Columbia to northern California. In Washington, there are five active volcanoes: Mt. Baker, Glacier Peak, Mt. Rainier, Mt. St. Helens, and Mt. Adams. The subduction-generated volcanic arc underlying the entire range began about 40 million years ago, although the modern topographic elevation of the range only developed during the last 10 to 5 million years. The active volcanoes that dominate the skyline in the Washington Cascades today began forming in the last 1 million years and are the most recent manifestations of volcanic arc activity.

Countless times over the past 40 million years, a new volcanic cone would appear somewhere in the Cascade Range, grow to its maximum size, and gradually lose its topographic distinction as the destructive processes of erosion exceeded the rate of volcanic reconstruction. Episodes of volcanic activity interspersed with intervals of sometimes thousands of years of inactivity occur during the life history of each Cascade volcano. Eventually the magma source is exhausted, and the inevitable downhill transfer of rock material disfigures and eliminates most if not all of the topographic evidence that a volcano ever existed at a particular location.

The volcanic arc that formed the Cascades is multifaceted, with variations in intensity and position over time, but the tectonic setting remains the same—a subduction zone. The eastward-moving Juan de Fuca Plate, all that remains of the older Farallon Plate, is still being subducted beneath the westward-moving North American Plate along the Cascadia subduction zone, located about 80 miles west of the present coast. Magma of basaltic composition continues to rise along the tensional fracture zones at the Juan de Fuca spreading ridge about 375 miles west of the coast.

Plate boundaries, like the subduction zone in western Washington, are notorious for creating conditions conducive to the formation of magma and chains of volcanoes. Oceanic crust containing water and soft sediment is pushed and pulled down along the subduction zone into the underlying mantle. About 60 miles beneath the surface, water and gases released from the subducting slab partially melt

These feeder dikes, exposed along WA 410 in the Cascade Range, were the conduits through which magma reached the flanks and summit of a huge volcano that existed 25 million years ago. Note the person standing at the base of the steeply dipping dike.

the overlying mantle. The basaltic magma rises buoyantly toward the surface, melting the overlying crust as well. That zone of melting falls directly below the chain of active volcanoes that crown the Cascade Range.

Today, the North American Plate grinds westerly over the Juan de Fuca Plate at the alarming rate of 1.5 inches every year, a very rapid movement by geological standards. Stress continues to build on crustal faults, and igneous activity continues to feed the major active volcanoes along the Cascade crest. Numerous temblors, steam venting from fumaroles in volcanic craters, and an occasional destructive earthquake or volcanic eruption remind residents that they are on the leading edge—the leading edge of a moving continent!

Columbia River Basalt

Beginning 17 million years ago in far southeastern Washington, the ground split apart as fractures opened up. Huge volumes of magma welled up and flooded southeastern Washington and northern Oregon with basaltic lava, now called the Columbia River Basalt. It buried remnants of the old North American continent, the southern edge of the Intermontane terranes, and volcanics from the ancestral Cascade Range—basically all of the older rocks that occupied eastern Washington. Together, the multiple layers of basalt are as much as 3 miles thick in the Tri-Cities area. The Columbia Plateau, with a few exceptions, is very flat, as one would expect on the surface of a solidified lava lake. Fortunately,

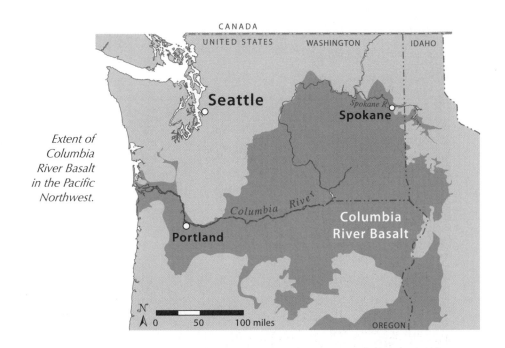

Extent of Columbia River Basalt in the Pacific Northwest.

The topography near Wenatchee changes abruptly between the flat Columbia Plateau and the Cascade Range in the distance.

Whitestone Rock, now exposed along a remote stretch of Lake Roosevelt about 14 miles upriver from the Sanpoil River, is some of the old rock buried by Columbia River Basalt. Dark-colored layers of basalt along the valley wall overlie the light-colored granites below. The granitic magma that cooled into Whitestone Rock intruded about 50 million years ago. For the next 35 million years the slow processes of uplift and erosion removed the overlying rocks, exposing for the first time the interior of the large batholith complex that forms many of the mountains in northeastern Washington. When massive flows of the Columbia River Basalt crept across eastern Washington 17 million years ago, they buried the edges of the mountains, including Whitestone Rock.

flood basalt fields of this magnitude are rare events because their extent, volume, rapid rates of extrusion, and effects on world climates would create more havoc than that portrayed in Hollywood disaster movies. The Columbia Plateau is the Earth's most recent example of such a megavolcanic event.

The Columbia River Basalt covers one-third of Washington State—over 81,000 square miles, mostly in Washington and Oregon, and extending into Idaho. Almost 90 percent of the lava erupted around 15 million years ago, in less than 1 million years, a blink of an eye in geologic time. Smaller

amounts continued to erupt in southeastern Washington until 6 million years ago. These immense lava flows did not come from a volcano, but from fissures many miles long. Similar but much smaller fissures form on the Big Island of Hawaii and in Iceland.

Such a huge volume of lava erupted in such a short period of time that individual flows were able to move 100 miles or farther from their fissure sources. They didn't cool and harden after moving a short distance because molten lava can move on the inside of a flow long after the outside of the flow has cooled. Some lava flowed all the way down the ancestral Columbia River to Vancouver, Washington, and Portland, Oregon, and into the Pacific Ocean. Rivers like the Columbia and Spokane were pushed to the edge of the plateau, where they flow today.

Pleistocene Ice Age

A trend toward cooler climates during the Cenozoic Era reached a critical condition about 2.5 million years ago at the beginning of the Pleistocene Epoch, the Earth's most recent ice age. About two dozen cycles of alternating glacial and interglacial conditions prevailed up to the present interglacial. World climates are precariously balanced, and changes in climate-influencing factors, such as variations in the Earth's orbit, changes in solar energy, or significant changes in the atmosphere, can trigger switches between interglacial and glacial conditions. Today, we are experiencing the warmest nonglacial interval of the past 2 million years. Alpine glaciers in the Cascades are particularly sensitive and are responding dramatically to climate change as their volumes decrease annually and smaller glaciers disappear.

The last glacial episode began about 32,000 years ago and ended by 12,000 years ago in Washington State. Extensions of the continental ice sheet that were over 1 mile thick flowed southward from Canada through the major valleys of northern Washington. Fingers of ice moved through the Puget Sound lowland and the valleys of the Okanogan, Sanpoil, Columbia, Colville, and Pend Oreille Rivers, leaving distinctive depositional and erosional features in their paths. The huge glaciers ground rock particles into smaller sizes and moved vast amounts of loose rock material southward.

Alpine glaciers decorated some of the higher peaks south of the ice sheet, as well as peaks protruding above the ice sheet in northern Washington. The ice disrupted stream

Curtis Glacier in the North Cascades, one of a number of glaciers on Mt. Shuksan's flanks, has retreated significantly since the mid-1800s, a trend followed by well over 90 percent of Earth's glaciers.

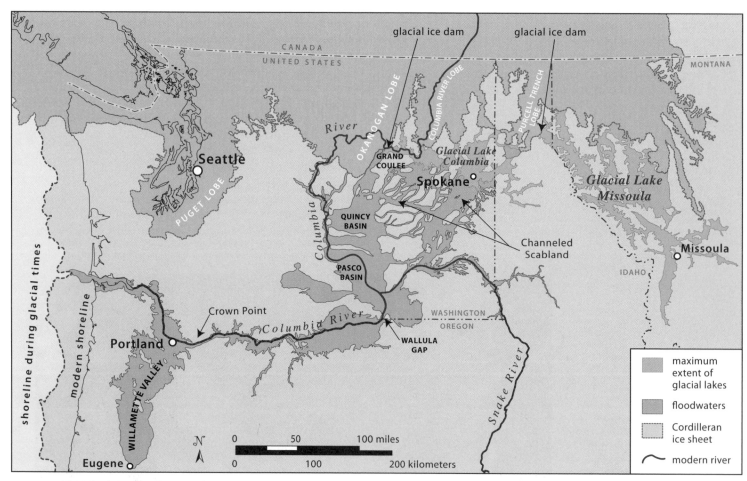

Glacial Lake Missoula in northwestern Montana, the ice dam in northern Idaho, the Channeled Scabland in eastern Washington, and other areas impacted by the Missoula Floods. —Modified from Ice Age Floods Institute, 2010; Miller, Roadside Geology of Oregon, 2014

drainages, and some major streams, such as the Pend Oreille and Colville, completely changed directions.

A tongue of ice in the Purcell Trench in northern Idaho blocked the Clark Fork River and impounded a huge lake, called Glacial Lake Missoula, which inundated the valleys of western Montana. The ice dam formed and collapsed dozens and perhaps hundreds of times. Each time the ice dam burst, hundreds of cubic miles of water careened through the Spokane River valley, across eastern Washington, and exited along the lower Columbia River into the Pacific Ocean in the span of just a few days. The floods left behind a torn-up landscape known as the Channeled Scabland. The floods redistributed glacial sediment and flushed huge amounts of finer sediment into the Willamette Valley in Oregon and eventually into the Pacific Ocean. Glacial Lake Missoula and its outburst floods ceased when the Cordilleran ice sheet retreated north of the Clark Fork River in northern Idaho.

Artist's rendition of an ice dam collapse releasing the pent-up waters of massive Glacial Lake Missoula.
—Drawing by Judy McMillan, Eastern Washington University

The Modern Landscape

The formidable Cascade Range, a barrier creating a classic rain shadow, separates the humid maritime climate of western Washington from the drier, arid to semiarid eastern part. Compared to areas east of the mountains, where only older remnants of a former violent tectonic and volcanic past remain, areas in the Cascades and westward are still in the throes of mountain building, making the Cascade and Coast Ranges some of the youngest mountain ranges in western North America. Volcanic eruptions, earthquakes, tsunamis, and landslides are some of the hazards of living in Washington.

Many young rock deposits in the state are especially prone to landslides. Unconsolidated volcanic and glacial deposits on the steep ridges and valley walls of western Washington can collapse without warning, especially when saturated with water. The 2014 slide in Oso was a sad reminder of this hazard.

The Jackson Springs (Miles) landslide in eastern Washington occurred catastrophically in March of 1969 when a steep bank of Glacial Lake Columbia sediments collapsed and completely blocked the flow of the Spokane River for a few hours. Fortunately, unlike the Oso landslide, no loss of life occurred. —Photo by Dale Stradling

The sudden failure of Pleistocene sediments in the hillside above the town of Oso in March of 2014 buried forty-five houses and killed forty-three people, one of the worst natural disasters to occur in the state during recorded history. —Photo by Barbara Kiver

OKANOGAN HIGHLANDS
Ancient Rock of Northeast Washington

The oldest rocks in Washington State, part of the ancient North American craton, are in the Okanogan Highlands, a logical place to begin our geological tour of the state. When the Intermontane terranes accreted to the old continent about 160 million years ago, a band of ocean floor rock—the Kootenay Arc—was squeezed between the colliding landmasses.

This part of the Northern Rockies is dominated by north-trending mountains and valleys. South-flowing glaciers up to 1 mile thick filled the valleys during the ice age. The glaciers removed much of the deeply weathered bedrock, so where it still exists, usually on the tops of high peaks, we know the mountains protruded above the glacial ice.

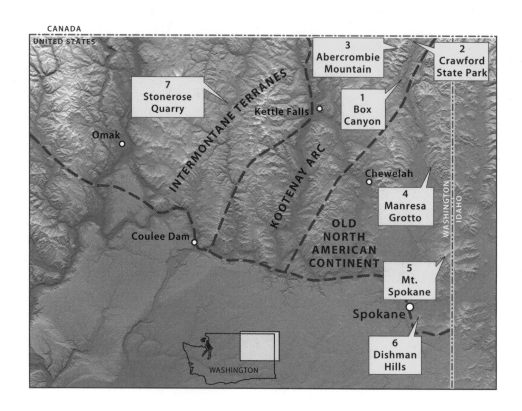

Typical mountain topography in northeastern Washington near the town of Hunters on Lake Roosevelt. The ice sheet, which flowed from north (upper left) to south (lower right) overrode and rounded these mountains, which are made of early Paleozoic bedrock. Note the streamlined glacial ridges beyond the lake in the foreground and middle ground.

Deep weathering on mountain summits, such as here at Gleason Mountain in Pend Oreille County, produces core-stones, which formed in place as air and water weathered and eroded the rock along fractures. These boulder features, called tors, help delineate the maximum height of Pleistocene ice in the Pend Oreille River valley and elsewhere in northeastern Washington. Glaciers could not have overrun such areas without removing the weathered bedrock or depositing transported rock material. —Photo by Dale Stradling

1. BOX CANYON

Backward-Flowing River Meets Limestone Ridge

The Pend Oreille River presents an unusual anomaly in northeastern Washington. It flows south from Canada into Lake Pend Oreille in northern Idaho, but instead of continuing south to the Columbia River like many other rivers do, it turns northward and flows back to Canada, joining the south-flowing Columbia River there. The Pend Oreille channel also narrows northward, from 1 to 3 miles wide near Newport to less than 100 feet wide at Box Canyon, to 15 feet in Z Canyon, which is now drowned by the reservoir behind Boundary Dam near the Canadian border. The width of a river channel normally increases downstream as the flow increases, but the Pend Oreille River channel does the opposite. The modern river is flowing down a drainage that was originally eroded by a stream flowing in the other direction.

Massive glaciers that occupied the area during the ice age are likely responsible for the change in direction.

Perhaps a stagnant or detached mass of glacier ice temporarily blocked the Pend Oreille River valley, forcing water to drain northward. Perhaps the weight of the massive glacier pushed down the crust, aiding the northward drainage.

The newly northward-flowing river cut the narrow Box Canyon through metamorphosed limestone, a carbonate rock. A turnout just south of Box Canyon Dam, north of Ione on WA 31, provides a spectacular view of the light-colored Metaline Limestone, which was deposited in the sea along the coast of the old North American continent about 520 to 490 million years ago, in Cambrian time. Invertebrates, such as trilobites and shelled brachiopods, inhabited the warm sea. Trilobite fossils help constrain the age of the Metaline Limestone, and deformed samples have been used to determine directions of tectonic compression.

Narrow canyons through tough bedrock, such as Box Canyon shown here, are ideal sites for the construction of hydroelectric dams.

A trilobite fossil in the Metaline Limestone.
—Photo by Larry Conboy, Eastern Washington University

2. CRAWFORD STATE PARK
Gardner Cave in Metaline Limestone

Less than a half mile from Canada off County Road 2975 about 12 miles north of Metaline is Crawford State Park, home to Washington's third longest limestone cavern and its only limestone show cave. As North America continued its eastward drift some 500 million years ago, warm, shallow seas allowed lime-secreting invertebrates such as trilobites to prosper along the west coast. Calcium-rich shells, shell fragments, and limy muds contributed to what later became the 1-mile-thick Metaline Limestone. Mountain building lifted the rock above sea level about 100 million years ago, allowing slightly acidic groundwater to slowly dissolve the limestone. Tiny holes developed into larger openings over many thousands of years. Groundwater flow abandoned the Gardner Cave section at least 20,000 years ago, and the air-filled chambers became sites of calcite deposition. Calcium-rich water seeped into the cave and deposited stalactites (hanging downward from the cave roof), ribbon stalactites or curtain formations, stalagmites (pointing upward toward the ceiling), a massive dome, and other forms.

Active cavern formation continues today in newer underground passages that are much too small to accommodate a human. Water pours into a sinkhole, also called a swallow hole, alongside the Frisco Standard Road (County Road 2975) about 0.7 mile to the southeast of the Gardner Cave parking lot. The water emerges with a roar from a small cave entrance near the Pend Oreille River that is not accessible to the public at this time.

Although miners originally came to this area for placer gold in stream gravels along the Pend Oreille River, the most significant economic deposit is lead and zinc ore and, to a much lesser extent, copper ore that mineralized in two zones of the Metaline Limestone. The rich metal-mining history of northeast Washington is reflected in the local town names at nearby Metaline and Metaline Falls. An underground lead-zinc operation, the Pend Oreille Mine, reopened in 2014 near Metaline Falls after being closed for maintenance since 2009.

Water flows down the outside of this massive column and adds a minute amount of calcite each year, but many thousands of years would be required to significantly change the diameter of the column. A large flowstone apron spreads laterally from the column base.

Rounded glacial erratics at 6,900 feet elevation litter the flanks of Abercrombie Mountain, indicating that ice reached this high during the last glacial episode. Ice in the valley was 1 mile thick here. Red backpack for scale.

The steep, pointed top of Hooknose, the 7,200-foot peak at the northeast end of the ridge of Abercombie Mountain, protruded above the valley ice lobe in glacial times. Pewee Falls cascades more than 200 feet over Cambrian limestone into Boundary Reservoir in the Pend Oreille River valley.
—Photo by Jeff Tetrick, US Department of Transportation

3. ABERCROMBIE MOUNTAIN
A Nunatak Peak

East of the Columbia River in northeastern Washington and British Columbia is a high range of mountains, the Selkirks. Towering over the US portion is Abercrombie Mountain, the second highest peak in eastern Washington, at 7,310 feet. The east face of Abercrombie looms more than 5,000 feet above the Pend Oreille River. A well-maintained Colville National Forest trail, part of the Pacific Northwest National Scenic Trail, leads to its summit and is accessible from the Pend Oreille Valley (east side) and the Columbia River Valley (west side).

During the last glaciation, ice flowed south from Canada and buried the northern mountains except for peaks, called nunataks, that were tall enough to poke through the ice sheet. Only the upper 400 feet of Abercrombie Mountain was ice-free, although the cirque on its north side hosted an alpine glacier. A view from its summit about 15,000 years ago would have resembled parts of Antarctica today.

The Addy Quartzite, a metamorphosed sandstone, forms the top of the mountain and is thousands of feet thick. The sand and overlying sediments were deposited in a sea to the west of the continent that persisted for hundreds of millions of years from late Proterozoic to Paleozoic time. The upper section of the Addy Quartzite contains early trilobites and other Cambrian-age marine fossils that appeared, geologically speaking, very suddenly at the beginning of the Paleozoic Era about 540 million years ago. The lower section is devoid of fossils, suggesting that this massive sand unit was being deposited during the transition from Precambrian to Paleozoic time, when hard-shelled fossils first began to appear in the geologic record.

4. MANRESA GROTTO
Rock Shelter in Conglomerate

On the east side of the Pend Oreille River just north of the Kalispel Tribal Headquarters is a unique cavelike opening in the bedrock wall of the valley. During the Pleistocene ice age, the river was blocked to the north by the retreating glacier, and Glacial Lake Clark formed in the valley to the south. Wind-driven waves crashed into the rock shore and preferentially enlarged a weak zone in the bedrock, forming a cavity about 60 feet long and 25 feet high and wide. Known as Manresa Grotto, the rock shelter was used by early missionaries as far back as 1841 and is still used today by the Kalispel Indians for certain ceremonies and tribal meetings.

The bedrock hardened from thick alluvial fans deposited along the sides of a valley that began forming about 50 million years ago, when an episode of volcanism and a change in tectonic plate directions in Washington stretched the crust. Blocks of rock dropped down along faults, forming valleys called grabens that filled with sediments. Groundwater seeping through the pile of sediments deposited calcium that bound the sand and gravels together into hard sedimentary rocks. Here in the Pend Oreille graben, the collection of sandstone, conglomerate, shale, and coal is called the Tiger Formation. Lowland swamp deposits in the valleys became coal. Rounded pebbles and cobbles in the conglomerate walls of the cave record a debris flow, a mixture of saturated sediment that flowed into the developing fault valley as a thick, viscous mass.

Manresa Grotto was excavated by waves along the shore of a large ice age lake impounded by the retreating Pend Oreille Valley ice lobe.

The conglomerate of the Eocene-age Tiger Formation contains pebble- to cobble-size rocks deposited by streams and debris flows that washed into the Pend Oreille graben. Image is about 3 feet across.

5. MT. SPOKANE
Dome of Granite

Just north of Spokane and visible from many tens of miles away is Mt. Spokane, the heart of Washington's largest state park. Its 5,883-foot-high elevation places it climatologically somewhere in northern Canada, much to the delight of ski enthusiasts. Spokane residents can live in a relatively moderate climate 4,000 feet lower and drive uphill for winter

Ski runs are visible as white streaks on the east flank of Mt. Spokane.
—Photo by Meghan Lunney

recreation. Mt. Spokane, as well as many other high, unglaciated peaks in northeastern Washington, features large fields of detached rock called felsenmeer, which means "sea of rocks" in German. These rock fields are not associated with rock falls or slides. Instead, they formed during the ice age climate when freeze-thaw cycles were more numerous and water freezing in rock crevices pried chunks of rock loose. The process continues today but with much less intensity than during ice age conditions.

The rocks in the felsenmeer are dominated by two granites. An older granite cooled from magma associated with a subduction zone some 100 million years ago. Later, crustal stretching released pressure on the Earth's crust and allowed a second pulse of magma to intrude about 50 million years ago. These rocks cooled at great depths in the Earth but were moved upward in the crust along low-angle faults. The granite is associated with highly metamorphosed rock and flanked by rocks deformed by the low-angle fault, forming what geologists call a core complex, or dome structure. Mt. Spokane is part of the Spokane Dome, discussed on the facing page.

Rock fields, like this one on the northwest flank of Mt. Spokane, formed on unglaciated peaks, mostly during the ice age.

6. DISHMAN HILLS
Gneiss of the Spokane Dome

The Dishman Hills, a rugged extension of resistant metamorphic and granitic rock, project into the heart of the Spokane Valley. Trails wind through rocky outcrops and ponderosa pine forest within the small Dishman Hills Natural Area, which is surrounded by urban development. Some trails follow northwest-trending valleys along a system of similarly oriented faults. The rocks at Dishman Hills were deeply buried in the Earth's crust and subjected to intense pressures and temperatures as the collision of numerous terranes thickened the crust during the subduction of the Farallon Plate 180 to 60 million years ago. The Precambrian-age Belt Supergroup was intensely metamorphosed and recrystallized into gneiss, a rock with a layered appearance, and pulses of granitic magma intruded it. When the plate motion shifted about 55 million years ago, the thick crust became unstable and stretched apart, allowing hot rock from deep in the crust to rise upward. The rocks are now exposed at the surface in a number of domal uplifts in northeastern Washington. The Spokane Dome, which is many tens of miles across, with its south flank located in the Spokane Valley, is exposed in the Dishman Hills, where you can see blocks of gneiss that were assimilated by the granite when it was still molten.

When Glacial Lake Missoula in western Montana broke through its ice dam and swept across the Spokane Valley, the Dishman Hills received a direct hit as floodwaters hundreds of feet deep roared over the peninsula of rock numerous times during the last glacial episode. The floods scoured the Dishman Hills, eroding soil and loose rock and exposing bedrock.

Caro Cliff in the Dishman Hills exposes folded gneiss injected by granitic dikes.

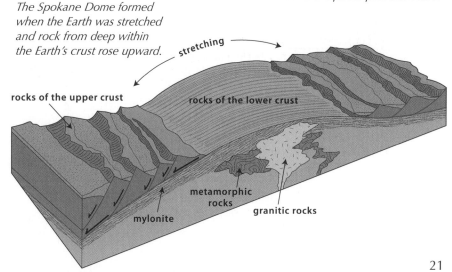

Rectangular blocks of gneiss were broken up and incorporated into granitic magma about 100 million years ago. A blue pen in the center of the photo provides scale.

The Spokane Dome formed when the Earth was stretched and rock from deep within the Earth's crust rose upward.

stretching

rocks of the upper crust

rocks of the lower crust

metamorphic rocks

granitic rocks

mylonite

7. STONEROSE QUARRY
Fossils from 50 Million Years Ago

In the town of Republic on State Highway 20 lies an amazing fossil quarry, where (for a small fee) you can split thin shale layers of the Klondike Mountain Formation along bedding layers to find fossils of leaves, flowers, seeds, and fruit that have not been exposed to sunlight for 50 million years. At the nearby Stonerose Interpretive Center, which displays a collection of the fossils, you can have your fossil identified before taking it home to add to your own collection.

At the beginning of Eocene time, a change in the direction of plate movement produced crustal tension and magma formation in the western United States. Large volcanic centers erupted lava and ash that covered the landscape. Blocks of Earth's crust dropped down along faults to form valley troughs called grabens, including the Republic graben in north-central Washington. The lowest part of the valley contained a 20-mile-long lake on the bottom of which accumulated an abundance of plant materials and dead insects. From time to time, fine volcanic ash settled to the lake bottom and buried the organic layers, preserving exquisite details of these long-dead life-forms.

Some of the fossil plants are now extinct, but their modern counterparts are climate sensitive, allowing scientists to reconstruct the physical environment of this part of Washington about 50 million years ago. The climate was much warmer than now because of unusually high atmospheric carbon dioxide levels. Also, the Cascade Range and Olympic Mountains did not exist at that time, so warm Pacific air could penetrate deeply into the continental interior. Palm, banana, cocoa, maple, and oak inhabited the hardwood forest, which had a much greater plant diversity than today's mostly coniferous forest.

The unusually warm interval in early Eocene time represents the most recent time besides the present when atmospheric carbon dioxide spiked to such a high level. Carbon dioxide levels rose extremely rapidly then, although not as quickly as is presently occurring. Volcanic emissions, tectonic rifting, a comet impact, and the rapid release of methane trapped on the ocean floor may have helped account for the warm climate.

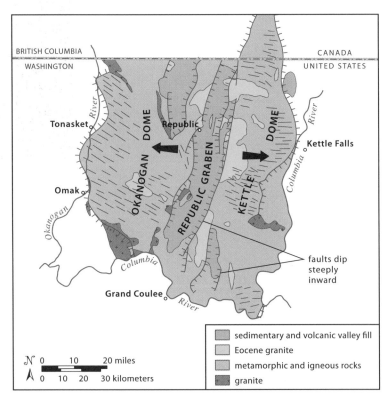

Sediments and volcanic material filled the down-dropped Republic graben in north-central Washington about 50 million years ago as the Okanogan and Kettle Domes were uplifted. —Modified from Alt and Hyndman, 1995

A young geologist-in-training is standing by a layer displaying leaf and other fossils in the Stonerose fossil quarry. Tilted layers of the Klondike Mountain Formation are visible on the upper slope.

Fossilized cocoa tree flower (Florissantia quilchenensis) is not technically a rose but from the same family and is used as part of the logo for the Stonerose Interpretive Center. The flowers can be up to 1 to 2 inches across. —Photo courtesy of Stonerose Interpretive Center and Eocene Fossil Site

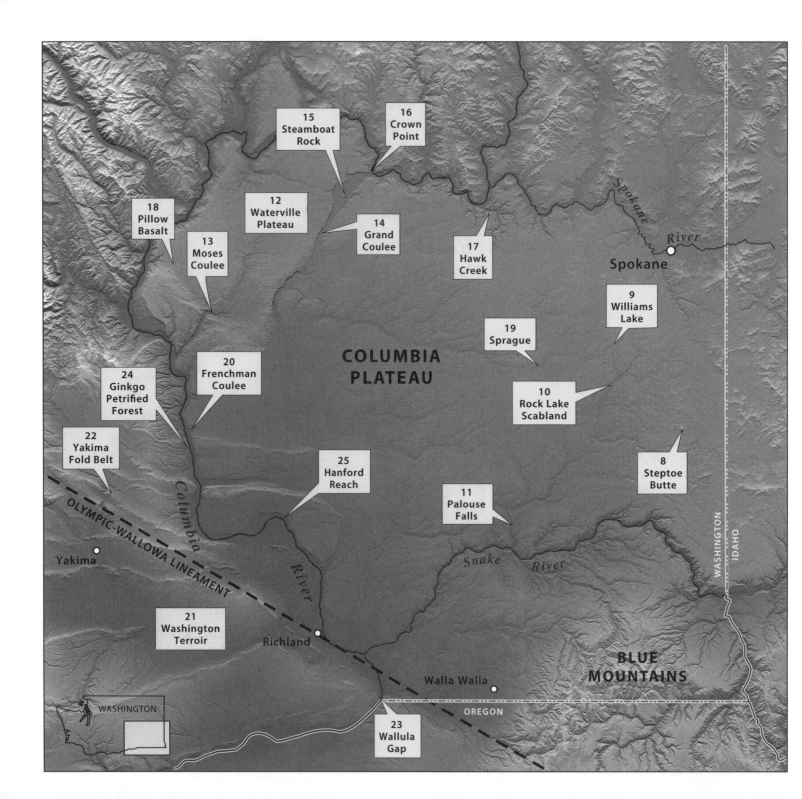

COLUMBIA PLATEAU
Massive Basalt Flows Ravaged by Catastrophic Ice Age Floods

The Columbia Plateau is easily delineated on a shaded-relief map of eastern Washington—it's the flat area surrounded by mountains. The voluminous flood basalts that began erupting 17 million years ago covered the area with layer after layer of lava, filling in the valleys and creating a fairly level plateau that covers over 7,000 square miles of the Pacific Northwest. Even more startling is that more than 60,000 cubic miles of lava erupted in less than 1 million years, a mere blink of the proverbial geologic eye. Some individual flows covered nearly the entire area in an estimated twenty-five days. The lava flows pushed the Columbia River and Spokane River to the flow edges, so they now follow the perimeter of the plateau on its north and northwest sides.

Although the region would seem to be simple layer-cake geology of massive lava flows, there are some complicating structural features. Wrinkle folds in central Washington, called the Yakima Fold Belt, extend eastward to form ridges and large structural basins in the basalt surface. The Pasco, Walla Walla, Quincy, and Ellensburg Basins trapped thick deposits of sediment. The structurally elevated Waterville Plateau occupies the northwest section, and the Blue Mountains poke up through the basalt and expose ancient bedrock just south of Clarkston.

The Okanogan ice lobe blocked the Columbia River during the last glacial episode and impounded Glacial Lake Columbia behind it. The southern position of the ice lobe, which also diverted Glacial Lake Missoula floodwater, dictated the place where the floods first carved Moses Coulee and later Grand Coulee. Repeated collapses of the ice dam that impounded Glacial Lake Missoula scoured the basalt, redistributed sediment over southeastern Washington, and flushed huge amounts of finer sediment down the Columbia River valley. The filling of the Spokane Valley with hundreds of feet of sediment created the world-famous Spokane aquifer, an ice age gift to the future states of Washington and Idaho. This huge volume of groundwater provides high-quality drinking and industrial water for the second largest city in Washington.

Deep gravel pits in the Spokane Valley expose the top of the Spokane aquifer.

View southeast over the scabland in the foreground to its junction with the wind-deposited Palouse soil in the midground, and Steptoe Butte in the distance.

8. STEPTOE BUTTE
An Isolated Rocky Mountain Peak

Steptoe Butte projects above the rolling hills of the Palouse at the eastern edge of the Columbia Plateau. Near here in 1858, the combined forces of the Spokane, Coeur d'Alene, and Palouse Tribes defeated the US Calvary under the command of Lieutenant Colonel Edward Steptoe. The butte is composed of 1.5-billion-year-old rocks of the old North American continent intruded by younger granites. About 15 million years ago lava flowed around the mountain, surrounding but not overrunning it with basalt. This geologic feature, so classically displayed here, is now known as a steptoe, and the term is applied to similar geological features around the world. Its uniqueness was further recognized in 1965 when it was declared a National Natural Landmark. Other steptoes near the edge of the Columbia Plateau, such as Kamiak Butte to the southeast, have a similar geologic story.

The access road to Steptoe Butte State Park winds three times around the conical butte before reaching the 3,612-foot-elevation summit, which stands 1,100 feet above its base. A lovely hotel erected at the top in 1888 burned to the ground in 1911 and was never rebuilt. On a clear day you can see over 100 miles to the Blue Mountains in Oregon and the Bitterroot Mountains in Idaho. The views of the rolling hills of the Palouse farmlands can only be duplicated from an aircraft.

View from the top of Steptoe Butte in northeastern Whitman County. Silt blown here during the ice age forms the rich agricultural lands in the surrounding Palouse Hills.

9. WILLIAMS LAKE
Dry Falls and Plunge Pool

A small acreage south of Cheney became the first new park established along the Ice Age Floods National Geologic Trail, an automobile trail that extends from the Glacial Lake Missoula basin to the Pacific Ocean. Williams Lake features one of the few dry falls and plunge pools in the Channeled Scabland that is accessible to the public and visible from a public road. To reach the park, take the Cheney Plaza Road south from Cheney, turn right (west) on Williams Lake Road, proceed north on Badger Lake Road for 0.3 mile, and park at the sharp right-angle turn.

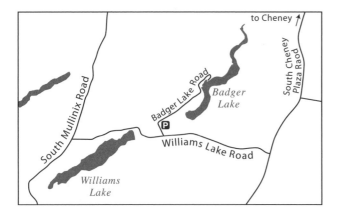

During episodes of glaciation an ice dam in northern Idaho impounded Glacial Lake Missoula, which contained as much as 530 cubic miles of water. Catastrophic failures of the ice dam sent floodwater a few hundred feet deep roaring at highway speeds across eastern Washington. The power and energy of the floods greatly exceeded that of the devastating March 2011 Japanese Tohoku tsunami. The run up from the tsunami in Japan was mostly from water depths of 30 feet or less. Missoula flood depths were commonly three to ten times deeper. The floodwater carved the Channeled Scabland in eastern Washington, a bizarre landscape of abandoned waterfalls with plunge pools, narrow canyons, coulees, and huge gravel bars. The canyons formed when the waterfalls eroded into the rock walls behind them.

A waterfall developed on the south end of Williams Lake and carved its way northward, forming the Williams Lake basin and plunge pool below today's dry falls. The last Missoula Flood was about 15,000 years ago—not very long ago in geologic time. With the exception of the vegetation, the area looks the same as it did when the last of the floodwater was gone.

The 100-foot-tall dry falls rim near Williams Lake wraps around the plunge pool.
—Photo by Barbara Kiver

10. ROCK LAKE SCABLAND
Flood-Torn Landscape with Silt Mounds

Tucked away in a remote area of the Channeled Scabland southeast of Sprague and just north of the little town of Ewan is the Rock Lake area, a monument to the impressive power of the ice age floods. Fishing and boat access is from WA 23 on the south side of Rock Lake. The Iron Horse/John Wayne Pioneer Trail along the old Milwaukee Railroad skirts the eastern shore, and a rails-to-trails extension to this area is in the planning stage.

Rock Lake was on the far eastern edge of the sea of turbulent floodwater that inundated a 100-mile-wide swath of eastern Washington numerous times during the Pleistocene ice age. Water preferentially excavated weak zones such as fractures or faults in the underlying basalt, deepening areas like the Rock Lake basin. At 400 feet deep, Rock Lake is the deepest natural lake in the Columbia Plateau. The lake bottom is about 1,000 feet below the rugged scabland topography on either side. Thus floodwater more than 1,000 feet deep briefly but effectively scoured the bedrock basin and surrounding area during flood events. A large cataract

formed near the mouth of Rock Lake and migrated 20 miles northward during the series of ice age floods. The water also cut the Bonnie Lake canyon immediately north of Rock Lake and Hole-in-the-Ground, a canyon between the two lakes.

The floodwater removed most of the area's cover of wind-blown silt, leaving only a few streamlined islands of silt. Following each Missoula Flood, sediment-covered surfaces were particularly vulnerable to wind erosion and supplied the bulk of the silt that makes up the Palouse soil of eastern Washington. Circular silt mounds, referred to as Mima mounds, are abundant in the Channeled Scabland. Their origin is a source of controversy, but wind likely played a major role in the formation of these silt mounds, which occur in thin soils overlying basalt or gravel. They probably formed by a different mechanism than the gravelly Mima mounds in western Washington (See Mima Mounds on page 94). According to J. R. Galm and K. McClure-Mentzer, some Mima mounds in eastern Washington bury Indian hearths that are less than five hundred years old.

This hill near the mouth of Rock Lake is a remnant of wind-deposited silt that was streamlined by the passage of giant floods from Glacial Lake Missoula.

Missoula floodwater exploded from the vertical-walled canyon at the mouth of the Rock Lake coulee and continued southward toward the camera location.

Aerial view of the rugged scabland along the east side of Rock Lake, where Missoula floodwaters ripped into the basalt, leaving numerous buttes, mesas, and closed depressions. The scattered beige circular features are Mima mounds composed of silt.

Palouse Falls and the lower canyon at sunrise at Palouse Falls State Park —Photo by Mark Kiver

11. PALOUSE FALLS
Waterfall Born of Violence

The ice age floods forever changed the course of the lower Palouse River in the southeastern part of the Columbia Plateau. The river abandons its large, wide valley, known as Washtucna Coulee, between the towns of Hooper and Washtucna and flows through an arrow-straight canyon toward Palouse Falls. Washtucna Coulee is a massive valley carved by the floodwaters, but apparently it wasn't big enough to hold it all. Floodwaters spilled over a drainage divide and followed a fracture in the basalt bedrock to the nearby Snake River. The thundering rush of Missoula floodwater dropping hundreds of feet into the Snake River canyon generated a waterfall that proceeded to cut headward nearly 6 miles to its present position at Palouse Falls, leaving a steep-walled canyon in its wake. The divide crossing was

The Palouse River followed a different course (dashed) to the Snake River prior to the ice age. During the Missoula Floods, floodwater followed fractures in the basalt (red lines), which allowed the Palouse River to find a shorter path to the Snake River. Palouse Falls started at the Snake River and retreated to its current location as the ice age floods ripped up the fractured basalt.

lowered enough by the floods that the Palouse River maintained its new course, abandoning forever its former westward channel along Washtucna Coulee.

The modern river spills 186 feet over Palouse Falls to a plunge pool below. The starkness of the basalt cliffs and the beauty of the waterfall have captured the attention of generations of people, including artists and photographers. The steepness and angularity of the topography and the thin to absent soils in the area are indicative of its geologically recent formation. In 2014 Palouse Falls was recognized as Washington State's official waterfall. Impressive as the waterfall is, especially during the spring runoff, it is a puny trickle compared to the 8-mile-wide stream of water that roared down the canyon and surrounding area at 50 or more miles per hour during larger outburst floods from Glacial Lake Missoula.

Palouse Canyon looking south from near Palouse Falls toward the Snake River, 6 miles away. Notice the zigzag pattern in the river, formed as the Missoula Floods preferentially carved out preexisting fractures in the basalt.

12. WATERVILLE PLATEAU
Deposits of a Continental Ice Sheet

The Waterville Plateau is an elevated area in the northwestern part of the Columbia Plateau that was uplifted along the Coulee monocline, a step-up type of fold in the plateau's basalt flows. The basalt layers of the Waterville Plateau are hundreds of feet above the level of their counterparts to the east and south. Agricultural lands and a few small towns dot the landscape.

During the ice age, the Okanogan lobe of the continental ice sheet covered the northern half of the Waterville Plateau. The Withrow Moraine, an elevated ridge of glacial debris, was deposited by the glacier when it was at its maximum southern extent about 18,000 years ago. Unusually large boulders and lumpy, irregular topography characterize the moraine. Rock debris of all sizes that was carried south by the moving glacial ice was deposited in one place as the ice melted at its southern edge. The Withrow Moraine is listed as one of Washington's National Natural Landmarks because it is the only terminal moraine on the Columbia Plateau and it helped shed light on the Missoula Flood story. The Okanogan Lobe blocked the Columbia River, forcing

The topographically impressive Withrow Moraine marks the southern extent of the massive glacier that occupied the Waterville Plateau at this location about 16,000 to 18,000 years ago. The moraine rises abruptly some 150 vertical feet just north of the small farming town of Withrow.

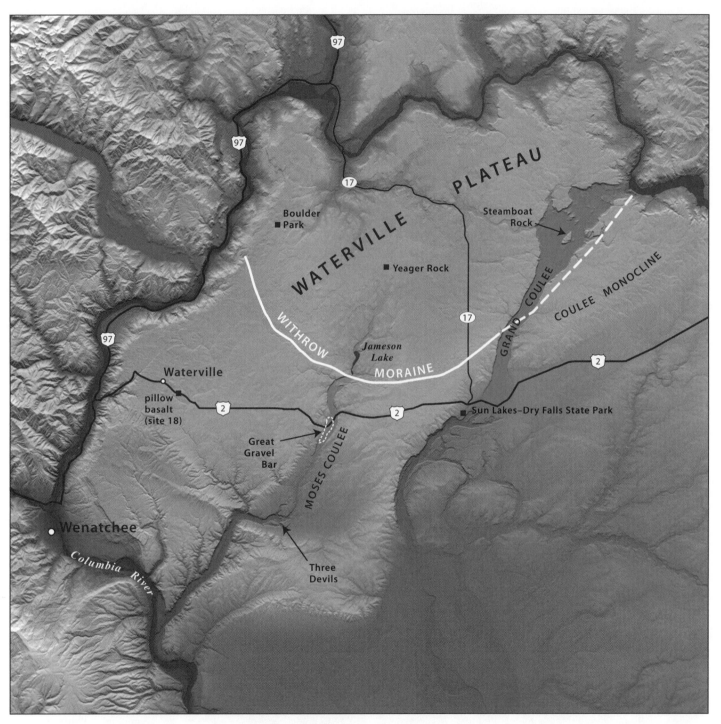

The northern half of the Waterville Plateau was glaciated, with glacial sediments burying much of Moses Coulee north of the Withrow Moraine.

floodwater to flow through and carve Moses Coulee and later, as the ice moved farther south, to cut the spectacular Grand Coulee. Bouldery soils north of the moraine are difficult to farm and less productive than the soil developed in the windblown silt that dominates to the south.

The area is a geological showcase for features formed by ice sheets. Huge boulder erratics, known locally as haystack rocks, are visible along the roadways on and north of the Withrow Moraine. Encased within the ice, they were transported by the glacier and left in place when the ice melted.

Yeager Rock between Mansfield and Sims Corner along WA 172 is one of the more accessible haystack rocks on the Waterville Plateau. Road builders wisely located the road south of this massive glacial erratic.

Glacially transported boulders, many the size of small houses, are abundant in the formerly glaciated area of the Waterville Plateau.

The Withrow moraine (not in photo) forms a natural dam in the upper section of Moses Coulee, impounding Jameson Lake. —Photo by Barbara Kiver

13. MOSES COULEE
Great Gravel Bar, Jameson Lake, and the Three Devils

The remote, 50-mile-long Moses Coulee, which runs south from west of Banks Lake to the Columbia River south of Wenatchee, is notched over 800 feet deep into basalt layers and contains McCartney Creek, a tiny intermittent stream. The valley is much too large to have been eroded by the trickle of water that exists there today. A large volume of water must have once flowed through its interior.

Between 28,000 and 20,000 years ago, during the early part of the last glacial cycle, the advancing Okanogan ice lobe blocked the Columbia River downstream of the present site of Grand Coulee Dam. The river, along with the huge

Missoula Floods, was forced to find a new outlet across the Waterville Plateau. The diverted water eroded Moses Coulee, a pathway that bypassed the massive Okanogan ice lobe and rejoined the Columbia River channel southeast of the present location of East Wenatchee.

Where the diverted water returned to the Columbia River, it spilled over the valley wall, generating a large waterfall, the location of which receded upstream rapidly as episodic floodwaters from Glacial Lake Missoula escaped through this newly formed outlet. Waterfall recession left canyon walls up to 800 feet tall that eventually extended 18 miles

GRAVEL   BAR

Aerial view looking southwest at the Great Gravel Bar, where US 2 crosses Moses Coulee. —Photo by Bruce Bjornstad

upstream to the now abandoned remnant of the waterfall in the Three Devils area. The spectacular dry falls area is where headward erosion of the canyon ceased.

The water also deposited an enormous gravel bar—250 feet tall and 3 miles long—on the west side of Moses Coulee upstream from the Three Devils. US 2 climbs the bar's ramplike surface out of the coulee and back to the upper surface of the Waterville Plateau. The gravel bar records the passing of floodwater hundreds of feet deep that descended the coulee and deposited more gravel with each passing flood. The gravel bar was deposited on an inside bend of the coulee, a place where the energy of the water current was reduced slightly. Individual flood events lasted for two or three days. Water not only filled the coulee but spilled out along its edges onto the plateau above. Because of its unusual size and geological significance, the Great Gravel Bar of Moses Coulee was listed as a National Natural Landmark in 1986.

About 20,000 years ago, the advancing Okanogan glacial lobe blocked Moses Coulee, and the Columbia River and Missoula Floods were forced to cut a new outlet farther east—Grand Coulee. The advancing glacier filled much of the upper part of Moses Coulee with sediment as far south as the Withrow Moraine, which impounds Jameson Lake, a natural body of water nestled in the coulee bottom and surrounded by basalt cliffs.

Canyon walls near the gravel bar in Moses Coulee expose some of the 15-million-year-old Columbia River Basalt flows. Weak zones in the basalt layers were preferentially enlarged during the Missoula Floods, and the attack by subsequent weathering processes helped form rock shelters, or overhangs, along the coulee wall.

Three Devils, a dry falls in Moses Coulee, separates the upper and lower coulee. Floodwater flowed from the top of the photo toward the bottom and spilled over the cliffs. This area is reached by a mostly gravel road between US 2 and WA 28.
—Photo by Bruce Bjornstad

37

14. GRAND COULEE
Dry Falls and the Coulee Monocline

Grand Coulee is about 50 miles long and is separated into an upper and lower coulee by a short flat stretch near Coulee City. The position of Grand Coulee was controlled in part by the position of the Coulee monocline, a single-flexure fold in rock that offset rock layers, placing them at different elevations. Inclined layers of Columbia River Basalt mark the limb of the monocline. More intense folding occurs to the south and west, where anticlines and synclines of the Yakima Fold Belt formed during an interval of north-south-directed crustal compression. In the Grand Coulee area, folding was more moderate, and the monoclinal fold trends northeast to southwest.

The ice dam across the Columbia River near Grand Coulee impounded Glacial Lake Columbia and forced the river and the episodic giant Missoula Floods to spill south over the Columbia Plateau, eroding Grand Coulee. Two enormous waterfalls formed along the ice edge: the upper falls where the Coulee monocline swings across the lower end of the Upper Grand Coulee, and a lower falls over the Soap Lake

anticline at the mouth of the Lower Grand Coulee. These waterfalls migrated upstream as the powerful currents continually undermined the lip of the falls. At the upper falls, floodwater raced down the slope of the monocline, forming a cataract over 2 miles wide. The waterfall maintained its clifflike form but migrated northward toward the spillway from Glacial Lake Columbia, leaving the deep canyon of the upper coulee in its wake.

The Lower Grand Coulee waterfall migrated north from near Soap Lake until it encountered the Coulee monocline. Here, the cutting of the Lower Grand Coulee followed the axis of the Coulee monocline, where the upturned edges of basalt along the monocline were weak, easily eroded zones. The floodwater encountered another bedrock structure near Park Lake and another recessional cataract formed, the present Dry Falls.

The topographic evidence is clear: Dry Falls would be the largest waterfall on Earth if a huge volume of water were added. From the Dry Falls Visitor Center along WA 17 and

Aerial view of the western section of Dry Falls, a complex of well-formed alcoves and plunge-pool lakes.

The alcoves and plunge-pool lakes as seen from near the Dry Falls Visitor Center. About 25 percent of the cataract complex is visible from this location; the other 75 percent lies out of view to the east.

View looking northwest over Park Lake in Sun Lakes–Dry Falls State Park at folded basalt layers in the Coulee monocline in the eroded west wall of the Lower Grand Coulee.

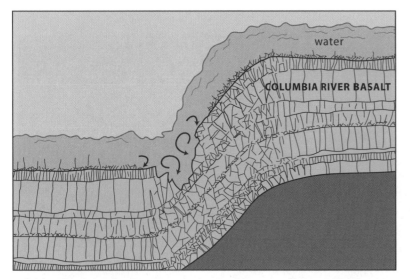

water

COLUMBIA RIVER BASALT

A fold in the Columbia River Basalt, the Coulee monocline, produced an elevation change that triggered waterfall development.

stretching 5 miles to the east to Don Paul Draw near the Pinto Ridge Road is a vertical to nearly vertical wall featuring a remarkable series of waterfall alcoves, plunge-pool lakes, and cliffs. The width of the floodwater surface that funneled into the Dry Falls complex and the Lower Grand Coulee exceeded 10 miles. To an observer near ground level during one of these floods, this would have appeared to be an immense river that stretched to the horizon. This fossil waterfall remains relatively unchanged since the last catastrophic flood swept over its lip sometime between 14,000 and 15,000 years ago.

The amount of water that flowed through Grand Coulee is impressive, but even more mind-boggling is the fact that each flood event was also simultaneously causing similar havoc in other scabland tracts 100 miles to the east. Stretching from the Washington-Idaho border to Grand Coulee, the episodic outburst floods were changing the landscape in a way that is not known to be duplicated at this scale anywhere else on our planet.

The deepest part of the Lower Grand Coulee contains a number of delightful lakes, including Park Lake at Sun Lakes –Dry Falls State Park, where an excellent cross-sectional view of the monocline occurs along the west wall. You can catch glimpses of the monocline at the mouth of the Upper Grand Coulee along WA 155, at Lake Lenore, and at other locations in the Lower Grand Coulee.

Columbia River

Grand Coulee Dam

Steamboat Rock

UPPER

GRAND COULEE

Sun Lakes– Dry Falls State Park

Coulee City

Dry Falls

LOWER GRAND COULEE

COULEE MONOCLINE

southern edge of ice lobe

The Coulee monocline was instrumental in the development and position of Grand Coulee.

15. STEAMBOAT ROCK
Remnant of a Vanished Waterfall

Like a huge ship on water, Steamboat Rock dominates the northern part of the Upper Grand Coulee. The layers of the Columbia River Basalt form the upper part of the isolated mesa, and the underlying granite of the Okanogan Highlands is well exposed at its base on the north end. During the ice age, the Okanogan ice lobe flowed south, blocking the Columbia River and diverting the entire river onto the surface of the Columbia Plateau. Outburst floods from Glacial Lake Missoula to the west followed the same course along the ice edge, wreaking havoc on the landscape and creating the 900-foot-deep canyon now known as Grand Coulee.

At the lower end of the Upper Grand Coulee, Missoula floodwater poured over the Coulee monocline, a folded section of the Columbia River Basalt, forming a waterfall several hundred feet high. Hydraulic forces maintained the vertical drop as the waterfall eroded its way upstream about 18 miles to the Steamboat Rock area, where it widened considerably upon encountering the hard underlying granite. The floodwater flow split, leaving Steamboat Rock, an island-like remnant. This type of feature is called a goat island after a similarly formed feature in upstate New York—Goat Island separates the American and Canadian sections of Niagara Falls.

View looking south over Steamboat Rock, a 900-foot-tall monolith of basalt in the Upper Grand Coulee. The peninsula that extends to the left from the base of the rock is made of granite that underlies the basalt.
—Photo by Bruce Bjornstad

Steamboat Rock, the large isolated mesa in the Upper Grand Coulee, was named by early settlers back when the coulee was dry. The construction of Dry Falls Dam at Coulee City created Banks Lake, and now Steamboat Rock looks even more like a ship.

STEAMBOAT ROCK

GRAND COULEE

View to the south of the Upper Grand Coulee, cut when the east side of the Okanogan ice lobe was situated near the site of the present dam, diverting the Columbia River and the massive Missoula Floods southward. The granitic rock (light rock at the edge of the dam and extending to lower right half of photo) is rounded by the ice and disappears southward beneath the edge of the thick, flat-lying basalt of the Columbia Plateau (visible in the distance). Crown Point is just off the photo to right.
—Photo by Paul Weis

16. CROWN POINT
Ice Dam at Grand Coulee Dam

Just north of the town of Grand Coulee off WA 174, the vista at Crown Point provides an amazing, airplane-like overview of the boundary between the basalt of the Columbia Plateau and the granitic rocks of the Okanogan Highlands. Grand Coulee Dam below the vista utilizes the tough granitic rock for its foundation. The old rock of the Okanogan Highlands disappears southward beneath the edge of the thick basalt of the Columbia Plateau.

The concrete structure of Grand Coulee Dam is relatively insignificant compared to the former ice dam that existed here. The Okanogan ice lobe flowed south, down the Okanogan Valley, and dammed the Columbia River, forming Glacial Lake Columbia, which flooded far upstream into the Spokane Valley and northern Idaho. The ice-polished surface of the granitic bedrock at the vista still shows scratch

marks left by debris embedded in the base of the ice lobe as it moved across this site. Large, detached boulders of basalt and granite that lie randomly on the surrounding hillsides were left by the retreating ice age glacier.

The Columbia River and Glacial Lake Columbia were swollen by melting glaciers during the warmer summer months, and water flowed southward along the eastern margin of the Okanogan ice lobe, initiating the channel that would ultimately become Grand Coulee, one of the most spectacular and accessible of the Channeled Scabland features. Episodic collapse in northern Idaho of another ice dam, the one holding in Glacial Lake Missoula, sent catastrophic floods of water across eastern Washington, including through the Grand Coulee floodway. After the ice melted, the Columbia River resumed its northerly path here.

17. HAWK CREEK
Sediments of Glacial Lake Columbia

The massive Grand Coulee Dam holds back Lake Roosevelt, a Columbia River reservoir some 150 miles long that extends from the dam to within a few miles of the international boundary. Today's impressive reservoir pales in size compared to its ice age predecessor, Glacial Lake Columbia. The Okanogan glacial lobe spilled into and completely blocked the Columbia River valley west of Grand Coulee, impounding a massive lake. Glacial Lake Columbia at its maximum elevation of about 2,400 feet was nearly 1,100 feet deeper than present-day Lake Roosevelt.

Hawk Creek Campground, part of Lake Roosevelt National Recreation Area, is at the mouth of Hawk Creek, which drains northward off the Columbia Plateau into Lake Roosevelt. The creek cut deeply through the thin-bedded to laminated, fine-grained sediments that accumulated in the relatively still waters of Glacial Lake Columbia. A close look at each bed reveals a darker zone, deposited in winter when fine mud settled to the lake bottom, overlain by a lighter zone, deposited during summer as upstream glaciers melted and released silt. Each dark and light layer together form an annual couplet, called a varve, that can be read like tree rings to determine the age of the lake. During the last glacial advance, Glacial Lake Columbia persisted for about 2,000 to 3,000 years, probably from about 18,000 to 15,000 years ago.

Every decade or so, many of the fine-grained varve sequences in Glacial Lake Columbia were interrupted by a layer of coarser material deposited by the entry of fast-flowing

Glacial Lake Columbia.

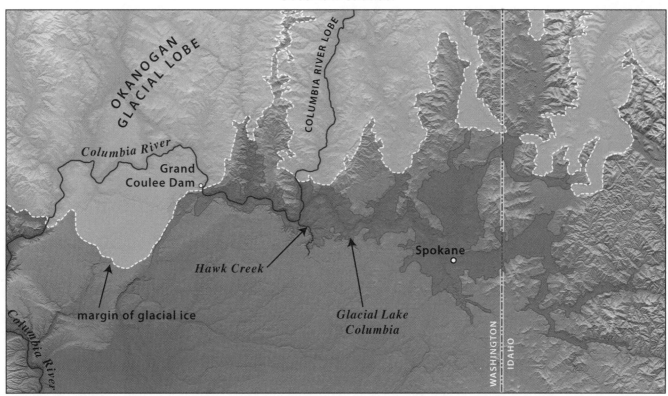

Lake sediments deposited in Glacial Lake Columbia often display varves, yearly couplets of dark winter sediment and light summer sediment. At least twenty-four years of sediment accumulation are shown in this small section of lake sediment.

water from the failure of another ice-dammed lake, Glacial Lake Missoula, in northern Idaho and northwestern Montana. Thus, the number of floods and how frequently they occurred have been determined in part from the varves in Glacial Lake Columbia sediments. At least twenty-seven outburst flood events and likely more are present in the Hawk Creek area. Other areas around Lake Roosevelt have evidence for eighty or more catastrophic flood events during the last glacial episode.

As the Okanogan ice lobe melted back to the north and lake levels lowered at the end of the ice age, the Columbia River and its tributaries began to cut into and through the thick section of lake sediment. Hawk Creek also cut into the hard granite and basalt buried below the lake sediment and left a delightful waterfall in what is now the National Park Service's Hawk Creek Campground.

Hawk Creek has cut through the thick lake-deposit valley fill and into the underlying hard layers of basalt, forming a falls at the head of Hawk Creek Campground.

The high, steep walls of poorly consolidated lake sediment close to Lake Roosevelt and along Hawk Creek are areas of very active landslides. Thousands of small to large landslides are located along the reservoir edge, with dozens of new slides occurring every year. When Hawk Creek cut through the lake sediments at the end of the ice age, it did not follow its buried, former channel, so now groundwater flowing through the sediment-filled, former channel contributes to the landslides. Groundwater is retained by the low-permeability silt-to-clay-size lake sediments, adding considerable weight to the sediment terraces and reducing friction between sediment grains. In the spring, steep slopes facing the reservoir are heavily loaded with groundwater replenished from rains and melting snow. Occasionally the load exceeds the strength of the slope, and lake sediments slide down.

Aerial view of the mouth of Hawk Creek Bay in Lake Roosevelt. The high level of the reservoir (1,289 feet) is marked by the well-delineated, lower edge of vegetation. Note the many landslide scarps in the Glacial Lake Columbia sediment on the valley walls.

As the ice dam that held Glacial Lake Columbia retreated and erosion in Grand Coulee proceeded, the lake achieved some temporary lower levels, leaving terraces to mark the former lake and river levels. Five levels appear in this aerial view taken over Fort Spokane near the Spokane-Columbia River confluence in the mid-1980s.

Closer view of a rounded pillow structure and elongated tubes. Diameter of the pillow is about 3 feet.

Sunset view of roadcut along US 2 between the towns of Waterville and Douglas. The lava entered a body of water, either a lake, pond, or river, from the right, forming long, inclined beds of tubes surrounded by fine particles of yellowish clay. The exposure is about 20 feet high. —Photo by Barbara Kiver

18. PILLOW BASALT NEAR WATERVILLE
A Classic Outcrop

East of Waterville along US 2 near milepost 154 is a roadcut that reveals what happens when lava flows into water. Molten rock and water don't mix. Instead, the magma cools into unusual features called pillow basalt, which is relatively common in the Columbia Plateau, but the Waterville exposures are outstanding examples. As lava enters water, the lava surface is quenched into elongated tubelike forms. The liquid lava in the interior of the tube keeps pushing out through the tube wall like toothpaste emerging from its tube. When you see one of these tubes in cross section, it looks like a not-so-soft pillow. A fine, orange- and yellow-colored clay mineral called palagonite forms on the lava surface

from reactions with water, creating colorful exposures. Actively forming pillows have been observed by divers in Hawaii, where Kilauea's lava occasionally flows into the Pacific Ocean.

The relentless flow of northward- and westward-flowing lava on the Columbia Plateau eventually filled or displaced existing water bodies. The voluminous flows of Columbia River Basalt overwhelmed the preexisting topographic and hydrologic features, completely burying the landscape. Large rivers like the Spokane and parts of the Columbia River were pushed against the surrounding mountains, where they flow today.

19. SPRAGUE AREA
Spheroidal Weathering of Basalt

Pillow lava is not the only round feature you'll see in basalt. Spheroidal weathering of jointed basalt also forms rounded blocks. Molten lava in a thick flow that has stopped moving cools slowly, particularly near the flow base. A hot solid develops first and further cooling causes contraction cracks to form perpendicular to the cooling surface, usually the ground below the lava flow. If the cooling rate is similar throughout the lower section of the flow, known as the colonnade, five- to seven-sided, eye-catching polygonal cracks, called columnar jointing, propagate upward. In addition to the mostly vertical cooling fractures, horizontal fractures can also form and divide the columns into vertical, stacked blocks. Groundwater moving through the fractures chemically attacks the outer surface of the blocks. Minerals in the rock are weathered to clay that expands, forming concentric weathering rings that peel off like the layers of an onion. This striking weathering pattern is called spheroidal weathering and also forms in other jointed rocks, like granite. A good place to see these rounded blocks is along I-90 near the Sprague Lake Rest Area and along Lake Road, which is reached from First Street in downtown Sprague.

Close-up of spheroidal weathering along Lake Road. Three-inch lens cap for scale.

The columnar jointing in basalt along I-90 westbound, just west of the Sprague Lake Rest Area, is cut by horizontal fractures, forming stacked columns. This outcrop can be seen at a distance across I-90 from the eastbound rest area, or it can be reached from Lake Road, which runs along the top of these columns out of view to right.

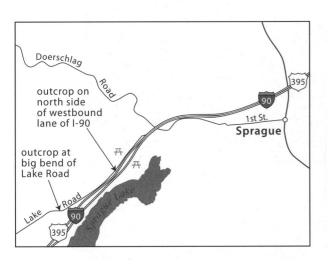

20. FRENCHMAN COULEE
Giant Columns of Basalt

Out of sight but almost within view of I-90, the state's major east-west highway, is a canyon enlarged by Missoula floodwater. Exit 143 leads to Silica Road and Vantage Road into Frenchman Coulee and dead ends on the east side of Lake Wanapum along the Columbia River. Water pouring out of Grand Coulee and other flood channels quickly filled the nearby Quincy Basin, and water escaped through three outlets, the southernmost of which is Frenchman Coulee. The ravaging floodwater spilled into the Columbia River valley far below and formed a group of waterfalls that receded upstream to form Frenchman Coulee. The vertical canyon walls in the main coulee expose layer upon layer of Columbia River Basalt, revealing details of the remarkable lava floods of about 15 million years ago.

The lower part of a thick lava flow is well insulated from the cooling effects of the atmosphere and cools very slowly compared to the upper part of the flow. The slowly cooling lava changes from a molten liquid into a very hot, mushy solid. Further cooling starting at the base of the flow causes the still hot but now brittle lava rock to contract and form regularly spaced contraction centers and radiating cracks that tend toward six-sided polygons. As cooling progresses, the polygonal cracking proceeds upward, forming columnar jointing. The column section, known as the colonnade, produces striking outcrops, often with nearly geometrically perfect shapes. Some of the columns in Frenchman Coulee have unusually large diameters, are 75 feet tall, and can easily be viewed from an automobile along the dead-end Frenchman Coulee Road. The upper part of the flow cools more quickly into an irregular blocky layer known as the entablature.

The Feathers in Frenchman Coulee are about 75 feet tall and up to 25 feet in diameter. They form a rock blade that separates Frenchman Coulee from a smaller coulee to the north.

21. WASHINGTON TERROIR
World-Class Wines in the Scabland Desert

Wine has been made in western Washington since the Hudson's Bay Company grew the first grapes in the early 1800s, but it was not until the early 1900s that irrigation systems allowed for extensive vineyards in eastern Washington. William B. Bridgeman, one of the pioneers of viticulture, moved from growing table grapes to European wine grapes in 1917 near the town of Sunnyside. Today, Washington is the second largest premium wine producer in the United States with more than 800 wineries and over 350 vineyards.

The term *terroir* is a French word used to describe physical factors, such as climate, soil, and geology, that combine to produce excellent grapes. Terroir contributes to Washington's success, too. Most of Washington's vineyards are located east of the Cascade Range in low-elevation parts of the Columbia Plateau, with many located in the Yakima Fold Belt and along the western slopes of the Blue Mountains. The summers are hot, with little rain and long daylight hours. The Cascade Range blocks westerly winds and creates a rain shadow, allowing many of the vineyards east of the Cascades to directly control the amount of water the vines receive. The Rocky Mountains along the eastern side of Washington block many of the coldest arctic storms moving down from Canada. Most of the vineyards in the Columbia Valley are in soils developed in windblown silts, deposits from the Missoula Floods, and volcanic-rich sediments from the Cascade Range. Newer vineyards are being planted in soils derived from basalt. These darker soils contain more iron and retain heat better. Although it seems odd that a dry area known as the scabland could host world-renowned vineyards, the moisture deficit allows the grapes to be stressed, producing juice with abundant sugars, perfect for wine.

Wine grapes with Columbia River Basalt in the background.
—Photo courtesy Jared Germain, Cascade Cliffs Vineyard & Winery

View northwest of Umtanum Ridge from WA 821 (Canyon Road) about 15.5 miles south of I-90, showing the layers of lava bent into a broad upward fold called an anticline.

22. YAKIMA FOLD BELT
Water Gaps through Folded Basalt

Between about 17 and 6 million years ago the enormous lava flows of the Columbia River Basalt filled the area between the Cascade Range and the Rocky Mountains, in Washington and Idaho, forming a series of huge lava lakes. Layer upon layer was added to the growing stack of basalt flows. In most areas, the layering is horizontal, and you can see this in cliff walls where erosion has cut into the basalt. However, in central Washington the lava layers are deformed into a series of ridges and valleys called the Yakima Fold Belt. The basalt layers resemble a throw rug on a smooth floor that has been crumpled by pushing from one side. The east-west orientation of the fold axes and steeper slopes on the north

flanks indicate that the tectonic force, or push, was directed from the south.

The exact geologic cause and timing of the folding is not known, although we know it began after most of the lava flows had been extruded and cooled. The oceanic plate that dives down into the subduction zone off the west coast moves in a northeast direction, at an angle to the coast. It is possible that the oblique movement between the oceanic and continental plates may have caused the folds. The fold belt is also crossed by the Olympic-Wallowa Lineament, a topographic feature that stretches diagonally across the state for 350 miles from the Olympic Mountains to the Wallowa

Basalt layers dip gently to the south on the south flank of the Frenchman Hills anticline. View looking east from Ginkgo Petrified Forest State Park.

The Columbia River cut this water gap, known as Sentinel Gap, through the Saddle Mountains anticline as the ridge was rising. View looking south from Ginkgo Petrified Forest State Park.

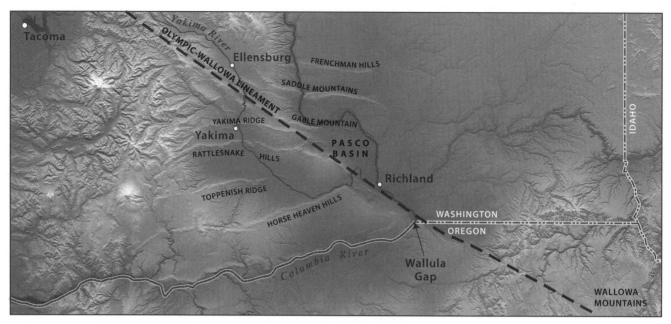

Ridges in the Yakima Fold Belt. The folds tend to bend in map view across the Olympic-Wallowa Lineament, suggesting that movement along that structure has bent and sheared the folds.

Mountains. The lineament has definitely influenced the orientation of the fold axes and intersecting faults, but whether it caused them is unknown.

Large rivers like the Columbia and the Yakima, which could erode at a rapid rate, were able to maintain their courses as the land beneath them uplifted and folded into ridges. The rivers cut impressive channels, known as water gaps, through the folded ridges and are the best examples in North America of antecedent rivers, ones whose paths existed prior to the formation of the structure they cross. The Yakima River canyon between Ellensburg and Yakima follows a course through three of these uplifted ridges and provides an opportunity to see numerous lava flows and their tectonic contortions. Particularly impressive is the canyon section through Umtanum Ridge, where you can see a textbook example of a cross section of an anticline from WA 821.

A wonderful overview of the Columbia River cutting through the Frenchman Hills anticline is seasonally available off I-90 at milepost 139 for westbound travelers about 3 miles east of Vantage. A short walk from the parking area to the viewpoint reveals the graceful upward curve of basalt layers to the north and south. The horizontal surface of Lake Wanapum provides a reference plane to judge the inclination of the lava layers on the fold flanks. A second anticline ridge in the distance to the south, the Saddle Mountains, is visible from the overview and the roadway. A prominent water gap there, Sentinel Gap, marks the continuation of the Columbia River as it marches southward across the central Washington landscape.

23. WALLULA GAP
Missoula Flood Bottleneck

The Columbia River flows through a relatively narrow water gap, called Wallula Gap, in the south-central part of the state about 6 miles north of Oregon. The river, as in other areas of the Yakima Fold Belt, defiantly cut through an imposing uplifted ridge—the Horse Heaven Hills anticline—which stands 1,200 feet above the river level. The powerful Columbia River only began cutting down into the fold during the last 5 million years, when the uplifting Cascade Range forced the Columbia River to shift course. Cross fractures along a fault that is part of the Olympic-Wallowa Lineament (OWL) north of Wallula Gap produced a weak zone in the basalt that the Columbia River exploited to cut Wallula Gap. The uplifted ridge, a fold in the 8.5-million-year-old Ice Harbor basalt, trends northwest from Wallula Gap to Benton City 33 miles away, where it bends sharply to the southwest upon intersecting the OWL. Other folds and faults associated with the Yakima Fold Belt also bend where they intersect the OWL. Feeder dikes for the Ice Harbor basalt, the conduits through which the magma reached the surface, have been offset by the OWL, indicating that at least some of the movement along the OWL is younger than the dikes.

During the Pleistocene ice age, when Glacial Lake Missoula sent hundreds of cubic miles of floodwater along complex pathways across eastern Washington, all routes led to the Columbia River and through Wallula Gap. The angry waters reached the bottleneck at Wallula Gap at a volume that was twice the capacity of the 1-mile-wide and 700-foot-deep gorge. Like a bathtub full of water draining through a small hole in the bottom, the water backed up to form Lake Lewis in the Pasco Basin, a temporary lake that required a week or more to drain. Fine sediments settled out of the lake water, covering a large area with silt after the last of Lake Lewis had drained through the gap. The wind blew the silt eastward, adding to the famous Palouse soil, which fuels the agricultural economy of much of eastern Washington.

View south into Wallula Gap at its narrowest point. The dashed line is at 1,200 feet elevation, the high point of floodwater during a large outburst flood from Glacial Lake Missoula. Downcutting by the Columbia River halted here in 1954 when McNary Dam impounded Lake Wallula.

24. GINKGO PETRIFIED FOREST
Semitropical Forest in the Desert

In the early 1930s road builders working on the cross-state highway US 10, the predecessor to I-90, encountered large quantities of petrified logs just north of Vantage. The discovery of rare ginkgo wood and the efforts of paleontology professor George Beck forced road engineers to move the roadway away from the fossil forest. Civilian Conservation Corps workers built trails, buildings, and protective structures around some of the logs, now featured in Ginkgo Petrified Forest State Park. In 1975 the legislature made petrified wood the state gem.

Following the massive extrusion of Columbia River Basalts about 15.5 million years ago, a quiet, lava-free interval of about 300,000 years allowed plants to revegetate the area. The warm, moist climate was similar to that of the southeastern United States today, and ponds and swamps covered much of the area. Floods or perhaps volcanic mudflows

from the Cascade Range, developing nearby, brought logs of higher-elevation trees like Douglas fir, hemlock, and spruce down and mixed them with low-elevation trees like walnut, sweet gum, swamp cypress, redwood, maple, elm, horse chestnut, and ginkgo. The fossil-rich deposit, known as the Vantage Formation, was sealed off by more lava flows. Groundwater moving through the ash-rich sediment slowly transferred silica to the tree cell walls, faithfully preserving the cell structure and converting the logs to petrified wood.

The ginkgo plant species evolved about 200 million years ago, coexisting with dinosaurs. The plant genus was thought to be extinct for millions of years until ginkgo trees were found growing in a remote area in China. Fossil leaf impressions have been found in many areas, but fossil ginkgo wood is extremely rare.

Small amounts of chemical impurities in microcrystalline quartz, known as chalcedony, produce a wide range of colors in petrified wood.

This petrified log near the park museum preserves the details of knots where branches emerged from the trunk.

The Columbia River in the Hanford Reach has eroded into soft sediments of the Ringold Formation, deposited in the Pasco Basin between 8 and 3 million years ago. Slope failure is common along these cliffs. —Photo by Bruce Bjornstad

25. HANFORD REACH
White Bluffs and Lake Lewis

During World War II, military planners in the top secret Manhattan Project were looking for a remote, relatively unpopulated area to build bombs. They found it in the Pasco Basin, a desert between the Saddle Mountains and Rattlesnake Hills, two major anticlines in the Yakima Fold Belt. The atomic bomb that was dropped on Nagasaki, Japan, in 1945 was built here using plutonium produced in the B Reactor. Eight other reactors were built here during the subsequent Cold War. The unfortunate legacy of this plutonium production is contaminated groundwater and many tons of nuclear and chemical wastes that have yet to be adequately cleaned up or safely stored. However, the security zone surrounding the contaminated area protected a unique desert environment, and in 2000 the area north of the Columbia River became Hanford Reach National Monument. This pristine shrub-steppe grassland is home to a number of species of

View from Wahluke Vista in Hanford Reach National Monument across the meandering Columbia River into the Hanford Nuclear Reservation, with the Rattlesnake Hills in the distance. Note the now inactive and cocooned nuclear reactor cores.

This ice-rafted erratic of argillite on the flanks of Rattlesnake Hills marks the location where an iceberg floating in Lake Lewis grounded and melted during one of the Missoula Floods. —Photo by Bruce Bjornstad

endangered plants and forty-eight rare or endangered species of animals. They are protected in the monument, which is managed by the US Fish and Wildlife Service. In addition, the Hanford Reach is the only nontidal, free-flowing stretch of the heavily dammed Columbia River. Over 80 percent of the native Chinook salmon in the Columbia River Basin utilize the Hanford Reach for spawning.

As the Yakima Folds formed the Pasco Basin, the sluggish ancestral Columbia River deposited fine sediment along its floodplain, a deposit now known as the Ringold Formation. At one time Ringold sediments filled the entire Pasco Basin. Erupting volcanoes to the west sent clouds of ash that settled on the floodplain, and layers of volcanic ash in the Ringold have radiometric dates from 8 to 3 million years ago. Regional uplift, along with rapid uplift of the Cascade Range about 4 to 3 million years ago, caused the Columbia River to cut downward. The newly energized river eroded most of the Ringold sediments, except

for the White Bluffs and Wahluke Slope in Hanford Reach National Monument and a few other small areas in the Pasco Basin.

Overlying the Ringold Formation are sands and gravels from the Missoula Floods, some of which are as young as 15,000 to 14,000 years old. Finer Pleistocene sediment in the basin was deposited each time floodwaters backed up behind Wallula Gap, the only hydraulic outlet in the basin. The temporary but huge body of water, known as Lake Lewis, lasted about a week before it drained.

Icebergs carried by the floodwaters blew aground along the edge of Lake Lewis, melted, and deposited their enclosed debris. Piles of gravel, called bergmounds, and large boulders, called erratics, can be found along the edges of the former lake and are made of granite, argillite, quartzite, and other rock types that were derived from somewhere to the north in northeastern Washington, northern Idaho, British Columbia, or western Montana.

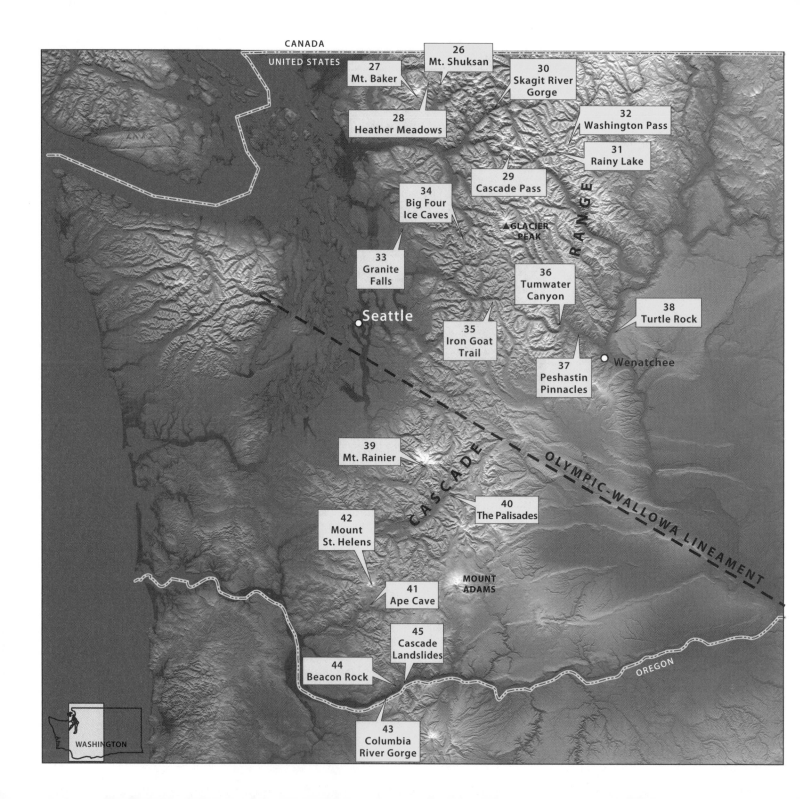

CANADA
UNITED STATES

26
Mt. Shuksan

27
Mt. Baker

30
Skagit River
Gorge

28
Heather Meadows

32
Washington Pass

31
Rainy Lake

29
Cascade Pass

34
Big Four
Ice Caves

▲ GLACIER
PEAK

R A N G E

33
Granite
Falls

36
Tumwater
Canyon

38
Turtle Rock

Seattle

35
Iron Goat
Trail

37
Peshastin
Pinnacles

Wenatchee

39
Mt. Rainier

40
The Palisades

42
Mount
St. Helens

CASCADE

OLYMPIC-WALLOWA LINEAMENT

41
Ape Cave

MOUNT
ADAMS

45
Cascade
Landslides

44
Beacon Rock

OREGON

WASHINGTON

43
Columbia
River Gorge

CASCADE RANGE
Washington's Volcanic Backbone

Extending from northern California to southern British Columbia, the Cascade Range is dominated by large, active volcanoes, the surface manifestation of the magmatic arc produced by the subduction of the Juan de Fuca oceanic plate beneath the North American continental plate. The heat and partial melting along the subduction zone generates magma that rises and erupts to form volcanoes.

The Cascades in Washington are divided by a major structural feature known as the Olympic-Wallowa Lineament (OWL), which slices diagonally northwest to southeast across the entire state. I-90 follows the lineament through the Cascades. North of the OWL, the Cascades have experienced more vertical uplift than their southern counterpart and thus are more deeply eroded by streams and glaciers.

Elevations north of the OWL are much higher, and rocks as old as 400 million years and perhaps older are exposed at the surface, although most are between 200 and 65 million years old. Rocks of similar ages to the south are deeply buried by up to 2 to 3 miles of volcanic materials, as generation after generation of volcanoes erupted over the past 40 million years. Volcanoes erupted in the northern Cascades, too, but more of the volcanics there have been eroded away.

The ice-sculpted North Cascades are sometimes called the American Alps. Today, the largest concentration of glaciers in the conterminous United States is located in the North Cascades, but like glaciers elsewhere in the world, they are rapidly disappearing.

Glacier Peak, a seldom seen volcano, is tucked away in a remote area of the Cascade Range.

26. MT. SHUKSAN FROM ARTIST POINT
Thrust Sheet from the Deep

WA 542 leads eastward from Bellingham and ascends to Artist Point, one of a handful of areas in the conterminous United States where you can easily experience a true alpine, arcticlike environment. That said, the nearby ski area holds the world record for snowfall, at 95 feet, and in some years the upper part of the road is never opened because of persistent deep snow.

Among the bewildering array of rugged peaks visible from Artist Point is 9,131-foot Mt. Shuksan, dominating the skyline to the east and one of the highest nonvolcanic peaks in the Cascades. The Shuksan Greenschist, metamorphosed ocean floor basalt, and the overlying metamorphosed mudstone and sandstone were part of the deep ocean floor about 160 million years ago. Compressive tectonic forces metamorphosed the ocean floor rocks, with basalt changing to greenschist and the mud and sand becoming phyllite and quartzite. The presence of unusual blue-colored metamorphic minerals in some of the Shuksan Greenschist indicates that the rocks were buried about 15 miles beneath the surface, placing them under very high pressure, but were uplifted rapidly before they were heated to the high temperatures found at that depth. Such a series of events is best achieved in a subduction zone.

The rocks of Mt. Shuksan are part of a large slab, one of many thrust sheets that were detached from the crust below and slid, one on top of the other, westward during collision with the North Cascade microcontinent. The Shuksan thrust sheet was the uppermost of these slabs that form the western North Cascades and nearby San Juan Islands.

A blue mineral in the Shuksan Greenschist indicates that the rock was subjected to very deep burial followed by relatively rapid uplift.
—Photo by Ned Brown

Glacier-clad Mt. Shuksan is 160-year-old metamorphosed ocean floor rocks, whereas the rocks of Artist Point, in the foreground, are lavas erupted about 300,000 years ago from a vent much older than the nearby Mt. Baker vent.

27. MOUNT BAKER
A Sleeping Volcano

Dominating the skyline in the northernmost Washington Cascades is the glacier-clad, 10,781-foot Mt. Baker, the latest of a number of cones that formed and were partly or mostly eroded away in the last few million years. The Mt. Baker cone, however, like Mount St. Helens to the south, is no more than 40,000 years old—very young by geological standards. Like the other tall volcanic peaks in the Cascades, Mt. Baker is a stratovolcano, formed by piling up layers of stiff pasty lava, mudflows, and pyroclastic debris, which is fragmental material blown from the vent. Such volcanic cones are typically steep sided and very vulnerable to debris avalanches and mudflows, the most dangerous hazards associated with this type of volcano. Volcanic gases venting from the volcano are acidic and cause minerals to break down rapidly into clays and other fine materials. When the clays are mixed with water and loose rock fragments, mudflows that originate on these steep slopes can extend tens of miles down the surrounding valleys.

The snow-filled summit crater formed about 15,000 to 12,000 years ago. The active Sherman Crater below the summit started with a bang about 6,500 years ago, and its the last major eruption was in 1843. An episode of increased heat release in Sherman Crater in 1975–1976 did not develop into a full-fledged eruption, although local residents nervously watched puffs of steam drift from its summit. However, the increased emission of volcanic gases and rapid melting of parts of the crater glacier provided scientists, including this book's senior author, an opportunity to delve into the workings of an active volcano.

Sherman Crater as it appeared in 1976 when increased volumes of hot volcanic gases poured from the crater vents. Note the large vent along the lake edge emitting a gas jet. Annual layers of snow (dark and light bands) in the crater glacier subsided into a fold as melting in the underlying ice cave passage increased.

Mt. Baker towers above the Puget Sound lowlands. Telephoto view from Hovander Homestead Park near Ferndale.

View southwest across the Bagley Lakes valley and Heather Meadows toward flat-topped Table Mountain and Kulshan Ridge.

28. HEATHER MEADOWS
Trail of Fire and Ice

Heather Meadows, near the east end of the Mount Baker Highway (WA 542), is a wonderland of scenery within the western North Cascades. Pleasant trails, including the interpretive Fire and Ice Trail, follow the deep Bagley Lakes valley, which glaciers scooped out and occupied in the not-too-distant past, perhaps as recently as 14,000 years ago. Glacial lakes, called tarns, are nestled in the valley bottom, which leads upward to a cirque basin and Herman Saddle, a col where glaciers have lowered a rock divide between adjacent glacial valleys.

About 1.1 million years ago the top of a now extinct volcano collapsed into its magma chamber, forming a huge depression much like that at Crater Lake, Oregon. Known as the Kulshan caldera, the crater later filled with over 3,000 feet of ashy tuff and lake deposits. About 310,000 years ago, a volcanic vent that pre-dates Mt. Baker sent andesite lava flows across the top of and into the former caldera, and they piled up against the north rim, becoming very thick. Columnar jointing formed in the lava as it cooled. Continental glaciers overran the solidified lava at least three times and etched out the Bagley Lakes valley between the lava flow and the caldera wall. The thick andesite is more resistant to erosion than the surrounding rocks, so glacial erosion of the Bagley Lakes cirque on the north and rapid erosion of the underlying Kulshan caldera volcanic sediments on the south by Swift Creek produced Kulshan Ridge and the narrow, plateau-like Table Mountain, both made of andesite. The former valley bottom is now a topographic high, an example of topographic inversion.

The rock-charged base of the glacier that crept down the Bagley Lakes valley scratched and grooved the 310,000-year-old Table Mountain andesite. Excellent cross sections of four- to seven-sided lava columns polished by the glacier are abundant along the Fire and Ice Trail. Many of the non-vertical columns of andesite lava along nearby WA 542 are part of this unusually thick, valley-filling flow. Because columns form perpendicular to the cooling surface, those that cooled along the former valley wall are inclined at various angles, including horizontal.

Glacial ice polished the tops of these columns of andesite bedrock along the Fire and Ice Trail at Heather Meadows. The asymmetric shape of the notch in the rock indicates that ice flowed from right to left. The blocks of columnar rocks on the left are steps installed along the trail.

29. CASCADE PASS
On the Edge of the Ice Age

A side road heading east from WA 20 in Marblemount leads up the Cascade River and into the rugged interior of the North Cascades. At the end of the 23-mile-long Cascade River Road is a large National Park Service trailhead parking area from which you can see precariously perched glaciers along a 4,000-foot-tall cliff that leads from the valley floor to the top of 8,100-foot Johannesburg Mountain. The ridge of debris below the small glacier terminus to the west of the parking lot formed during the Little Ice Age, a worldwide cold event that lasted from the thirteenth century to about 1850. The glacier here was bigger then and deposited debris at its snout to form this moraine. For a time, the glacier melted back as fast as it advanced, allowing the stationary ice front to drop most of its load in one place.

An 18-million-year-old dike exposed near the mountain base intrudes the 150-million-year-old metamorphic rock of Johannesburg Mountain and other nearby peaks. A dike forms when magma solidifies in a fracture. This dike, which is made of the granitic rock tonalite, is over 9 miles long, is 1 mile wide in places, and may or may not have fed volcanoes or other surface eruptions. You can follow the dike along the 3-mile-long trail to Cascade Pass. Views of more glaciers that occupy bowl-shaped depressions cut into the mountain flanks await the inquisitive hiker at Cascade Pass.

Hanging glaciers on the flank of Johannesburg Mountain.

A moraine, the ridge of debris in front of the snow, marks the former position of a larger glacier that existed at today's Cascade Pass Trailhead until the middle of the nineteenth century.

30. SKAGIT RIVER GORGE
Devil of a Canyon

WA 20, the North Cascades Highway, follows the Skagit River drainage through the center of North Cascades National Park. Between Newhalem and Ross Lake, the Skagit River flows through some very narrow sections of canyon that are as much as 1,500 feet deep and only a few tens of feet wide at the base. Narrow canyons like this have not existed long enough to be significantly widened by stream erosion. Skagit Gorge, a particularly deep, V-shaped canyon, has a high stream gradient and steep walls composed of very tough, resistant metamorphic rock.

Geologists have determined that the Skagit River upstream of Skagit Gorge formerly flowed northward into the Fraser River in Canada. During the last glacial episode, the Fraser ice lobe in nearby British Columbia crept down its valley and eventually blocked the northward flow of the Skagit River, creating a temporary lake. Water backed up deeply into the basin that now contains Ross Lake and eventually found a gap in the imposing Picket Range to the west. The overflowing water raced down the mountainside, notching into the hard metamorphic rock. Melting glaciers greatly elevated the water flow during the summer melt season, providing even more erosive power. The canyon deepened sufficiently that when the Fraser ice lobe retreated about 15,000 years ago, the Skagit River maintained its new course rather than returning to its old channel. Seattle City Light has taken advantage of three unusually narrow sections to construct hydroelectric dams that furnish most of the power for the state's largest city.

Nineteenth-century prospectors looking for valuable minerals upstream apparently had a difficult time following

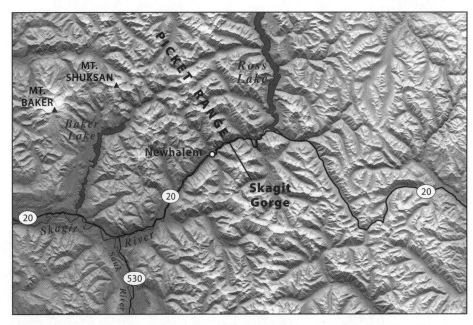

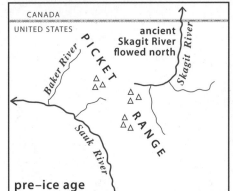

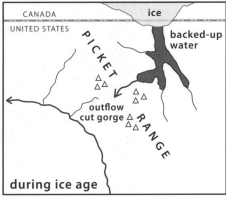

Prior to the ice age, the Skagit River drained to the north, but when its drainage was blocked by ice, it cut a narrow gorge through the Picket Range.

The roadcut across from the Diablo Lake Overlook exposes light-colored bands of 50-million-year-old granite that was intruded into the 90-million-year-old Skagit Gneiss Complex.

HAZARDOUS AREA
KEEP OUT

Diablo Dam towers over the narrow, vertical-walled canyon near the town of Diablo.

North of the national park visitor center in Newhalem is a breathtaking view of the jagged Picket Range and some of its glaciers.

the rugged canyon and named it Diablo. A WA 20 viewpoint for Diablo Lake, impounded by Diablo Dam, allows close inspection of the canyon's rock. The rock buttress at the parking lot may be a remnant of a former valley bottom that existed before Skagit Gorge was deepened by the water diverted across the mountains.

The bedrock is part of the Skagit Gneiss Complex, a group of metamorphic rocks whose minerals indicate the original rock was buried almost 20 miles below the Earth's surface. This section of the Cascades experienced very high metamorphic pressures and temperatures as large landmasses were added to the western edge of North America about 90 million years ago and the Earth's crust thickened. The original sedimentary and volcanic rock recrystallized into high-grade metamorphic rocks like schist and gneiss. Later, as the rock was uplifted and stretched, magma formed and was injected into the schist, becoming bands of light-colored granite.

Rainy Lake is the most easily accessed cirque in the North Cascades. Note the small glaciers and snowbanks occupying some of the higher cirques above.

31. RAINY LAKE
An Alpine Cirque

A short, 1-mile-long, paved path from the Rainy Pass Trailhead along WA 20 leads to a stunning lake nestled in a cirque, a bedrock bowl carved into the 90-million-year-old Black Peak batholith by a mountain glacier. A cirque is like a giant armchair with nearly vertical walls encircling a basin on three sides. Streams plunge precipitously down the rock walls, particularly in the early summer when snow is melting. Under favorable climatic conditions, like those 18,000 to 15,000 years ago, mountain glaciers send tongues of ice tens of miles down their valleys from the cirques where they begin. Strenuous hikes are usually required to visit a cirque, but because the North Cascades Highway is near its peak elevation at Rainy Pass, you can reach this cirque by a twenty-minute walk with only a minor elevation change.

North- to northeast-facing slopes or rock walls that provide shade are excellent places for cirques to form. Perennial snowbanks can form small basins that enlarge and trap more snow. If snow and ice thicken enough, they become a glacier that begins to move and further erodes the underlying basin. Any water flowing beneath the glacier enters cracks in the underlying rock, freezes, and pries more rock fragments loose as the basin enlarges.

The processes of cirque formation are most effective at snowline, the location and elevation where snow from the previous winter persists through the following summer melt season. The regional snowline today is currently well above the high peaks, although snowbanks and glaciers persist where they are protected by shade. The regional snowline during the last glacial episode was around 4,800 feet, the elevation of Rainy Lake. It is estimated that a 10- to 15-degree-Fahrenheit decrease in average air temperature in this alpine area would put us back into the next episode of ice age glaciation.

32. WASHINGTON PASS
Golden Granite of the North Cascades

For about four months each year a short side road is open from the North Cascades Highway to the Washington Pass Overlook at 5,550 feet elevation. During the remainder of the year Washington Pass and the viewpoint typically lie beneath 20 feet or more of snow, with 40 feet of snow in avalanche tracks. The avalanche-prone slopes above the North Cascades Highway are a major reason why the road is closed to travel during winter. One of the more dangerous sites lies below and south of the overlook along the long hairpin turn, where 60 feet or more of avalanche debris can accumulate.

From the overlook, an array of breathtaking peaks is visible along Kangaroo Ridge to the east and northward along the highway toward the glacier-carved Methow River valley. The peaks and pinnacles visible from here are composed of the eye-catching granite of the Golden Horn batholith. Granite is composed of mostly quartz and feldspar minerals, and some of the potassium-rich feldspars in the Golden Horn have a distinct pinkish or golden color. Solidified from a large body of magma, the batholith cooled about 2 miles below the Earth's surface about 50 million years ago. Erosion has since removed those 2 miles of overlying rock.

Monoliths such as the Liberty Bell, a favorite challenge for climbers, form in massive granite free of closely spaced fractures.

Pinnacles like these along Kangaroo Ridge, here covered by frost, form where closely spaced fractures occur in the granite bedrock. —Photo by Barbara Kiver

Potholes on granite surface a few feet below the walkway at Granite Falls. —Photo by Barbara Kiver

33. GRANITE FALLS
Potholes at a Knickpoint

Over time streams tend to smooth out their channel bottoms by meandering from side to side and removing the lumps and bumps. The South Fork of the Stillaguamish River has not had time to do that along its course through the Cascade Range. A parking area by the Mountain Loop Highway river crossing just north of the town of Granite Falls provides access to a stairway that leads down to a fish ladder and a remarkable series of falls on the river. Turbulent currents during floods swirl sediment around in small basins and grind potholes a few inches to a foot or more in diameter and depth. A visit during spring runoff is particularly exciting, although potholes in the rock are more visible during low water.

Solid bodies of rock resist the erosive power of water more than sediments or rocks with lots of fractures. Contrasts in the resistance of rock often produce significant vertical changes in channel elevations, called knickpoints, where waterfalls and rapids form. The resistant rock at Granite Falls is a small body of 48-million-year-old granite that remains stubbornly elevated above its weaker downstream neighbor, producing a substantial drop in vertical elevation.

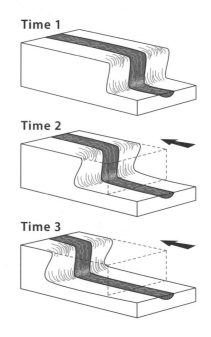

Over time, a waterfall moves upstream. This point where the elevation of a stream changes is called a knickpoint.

70

Upper waterfall at Granite Falls. —Photo by Barbara Kiver

34. BIG FOUR ICE CAVES
Look Out Below

During winter and spring the roar of tons of snow avalanching down the 4,000-foot, north flank of Big Four Mountain occasionally interrupts the stillness. Like distant thunder, the sound reverberates through the snowed-in valley of the South Fork of the Stillaguamish River about 20 miles upstream from Granite Falls. At the base of the mountain a 150-foot-tall snow cone is resupplied every year with fresh avalanche snow, enabling it to persist through the summer melt season. The north-facing slope limits the sun's rays and thereby reduces melting. Meltwater that flows between the snow and the bedrock forms large ice caves beneath the cone. The caves look very inviting but are extremely dangerous and should be viewed from a safe distance. Some who ventured into them were crushed to death by tons of collapsing ice.

The Darrington–Devils Mountain fault, a large, northwest-trending fault, slices the front of Big Four Mountain, accounting for the imposing vertical wall. The fault is a splinter off the Straight Creek fault, which cuts through the North Cascades from Snoqualmie Pass northward into British Columbia, where it is called the Fraser River fault. A shift in the oceanic plate movement to the northeast about 50 million years ago created this major north-south fault.

The snow cone along the front of Big Four Mountain forms by many tons of snow avalanching down the mountain front during winter. Note people standing near the edge of the snow and the small beginning of an ice cave that will enlarge during the summer melt season. Photo taken in early summer.

Entrance to the Big Four Ice Caves during the summer of 2013. If you can clearly read the danger sign, you are too close!

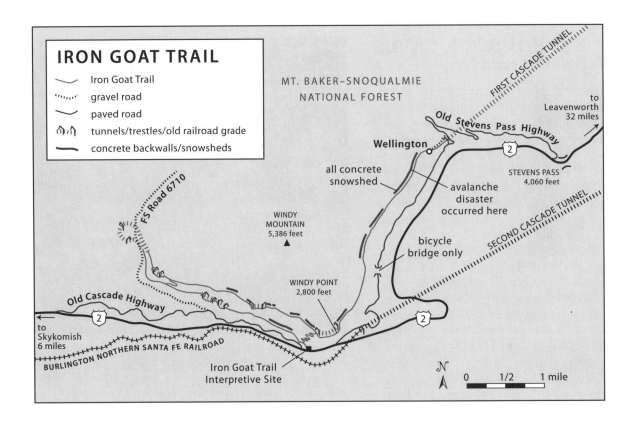

IRON GOAT TRAIL

IRON GOAT TRAIL

- ⌣ Iron Goat Trail
- ⋯ gravel road
- ⌣ paved road
- ⋏⋏ tunnels/trestles/old railroad grade
- ▬ concrete backwalls/snowsheds

MT. BAKER–SNOQUALMIE
NATIONAL FOREST

FIRST CASCADE TUNNEL

to
Leavenworth
32 miles

Old Stevens Pass Highway

Wellington

all concrete
snowshed

avalanche
disaster
occurred here

② STEVENS PASS
4,060 feet

FS Road 6710

WINDY
MOUNTAIN
5,386 feet
▲

bicycle
bridge only

SECOND CASCADE TUNNEL

WINDY POINT
2,800 feet

Old Cascade Highway

② ②

to
Skykomish
6 miles

BURLINGTON NORTHERN SANTA FE RAILROAD

Iron Goat Trail
Interpretive Site

N

0   1/2   1 mile

35. IRON GOAT TRAIL
America's Deadliest Avalanche

The most deadly snow avalanche and train disaster in North America occurred in the Cascade Range just west of Stevens Pass. A short spur road off the Old Cascade Highway provides access to the avalanche site, which is along the Iron Goat Trail. The trail utilizes the grade of the former Great Northern Railroad, which built the town of Wellington in 1893 and a 2-mile-long tunnel under Stevens Pass in 1900. The mountain slopes are very steep here, formed in tough granitic rock, part of the massive batholith complex that intruded the crustal rocks from 90 to 75 million years ago. The North Cascade microcontinent slid into place along a subduction zone beginning about 115 million years ago, triggering this major episode of intrusion and mountain building.

The railroad town of Wellington was located on a broad flat at the junction of two valleys high in the Cascades, in an area not expected to experience avalanches or landslides. The use of timber during construction and spark-caused fires from passing steam trains eventually removed much of the dense forest on the steep slopes above the town and nearby rail grade, thereby eliminating an important slope-stabilizing factor. An unusually heavy and persistent snow in late February of 1910 trapped two trains, one of which was a passenger train, for nearly a week at Wellington. On the night of March 1, 1910, warming temperatures destabilized the thick snowpack above, sending tons of snow into the edge of Wellington and carrying the two immobilized trains into the valley below. Ninety-six lives were lost.

View just west of Stevens Pass from US 2 showing the abandoned Great Northern rail bed (the dark horizontal line across the valley). Walls of the more than 100-year-old concrete snow sheds still stand, but the wooden roofs are gone and yearly snow avalanches once again cover the old train route.

Part of a half-mile-long concrete snow shed constructed in 1911 near the former railroad town of Wellington.

The 5,000-foot-tall, V-shaped canyon walls of the Wenatchee River in the lower Tumwater Canyon are cut in granitic rock of the Mt. Stuart batholith.
—Photo by Barbara Kiver

36. TUMWATER CANYON
A River Gone Wild

During spring snowmelt, the Wenatchee River in Tumwater Canyon dashes over rocks, white with flurry as it races to the Columbia River in Wenatchee. In Chinook jargon "Tumwater" alludes to fast water. The tough, resistant granitic rocks of the 95-million-year-old Mt. Stuart batholith maintain the steep cliffs of Icicle Ridge on the west and Tumwater Mountain on the east. The rock is slow to erode, causing the river to steepen its profile, dropping over 500 feet in 8 miles. Stevens Pass Highway (US 2) is squeezed into the bottom of the V-shaped canyon, providing access to numerous Forest Service turnouts but costing taxpayers each time maintenance crews must deal with a rockslide or washout.

Tributary streams like the Wenatchee are forever deepening their channels to reach the level of the larger body of water into which they flow. The huge Columbia River has been vigorously downcutting its channel, forcing the Wenatchee to keep up. The lower reach of the Wenatchee River was able to erode deeply through the weak sandstones and other sedimentary deposits that filled the Chiwaukum graben, a crustal block down-dropped 50 million years ago between the northwest-trending Leavenworth fault at the mouth of Tumwater Canyon and the Entiat fault farther north. Downcutting stalled west of the Leavenworth fault in the hard granitic rock of Tumwater Canyon, forcing the upper Wenatchee River into a steep gradient. Ice age glaciers never entered Tumwater Canyon as they did Icicle Creek in the next valley to the west, so the valley was not widened into a glaciated U-shape.

37. PESHASTIN PINNACLES
An Upstanding Group of Rocks

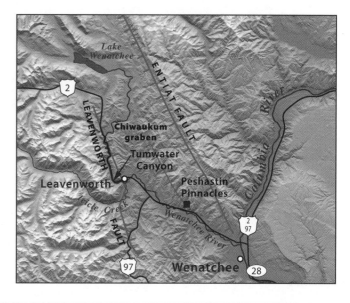

The sandstone slabs at Peshastin Pinnacles State Park southeast of Leavenworth along US 2 are a favorite challenge for rock climbers. Layers of the 46- to 40-million-year-old Chumstick Formation were originally deposited in horizontal or near-horizontal beds in the Chiwaukum graben, a large fault-bound basin, but are now steeply inclined at wild angles. The sediment pile that accumulated in the subsiding graben is estimated to be 6 miles thick.

The North American Plate has been converging with plates to its west for millions of years, but the plate motion shifted direction about 55 million years ago, moving to the northeast rather than directly eastward. The shearing motion generated the Straight Creek fault and its Canadian extension, the Fraser River fault, a strike-slip fault in which the block on the west side slides north relative to the block on the east side. The western Cascade block moved northward about 70 miles and caused the crust to stretch to the east of the fault, producing a number of normal faults. The Chiwaukum graben dropped down between the Leavenworth and Entiat faults. Topographically high areas on either side of the newly formed basin shed gravel, sand, and mud-size debris. Continued movement along the faults tipped and folded the layers, and some of them reached completely vertical orientations.

Tilted layers of the Chumstick Formation in Peshastin Pinnacles State Park.

75

38. TURTLE ROCK
Large Landslide Shaped by Ice Age Floods

Turtle Rock is a large island of Swakane Biotite Gneiss in Lake Entiat, a reservoir along the Columbia River impounded behind Rocky Reach Dam north of Wenatchee. You can contemplate the amazing sequence of events that formed Turtle Rock from a paved parking area with a Lake Entiat sign between mileposts 134 and 135 on the west side of US 2 north of the island. The Columbia River here flows along the boundary between the Cascade Range to the west and the Waterville Plateau section of the Columbia Plateau to the east. About 15 million years ago the massive flows of the Columbia River Basalt forced the river westward to near its present position. Isolated remnants of basalt overlie the Swakane Biotite Gneiss high on the west side of the valley, indicating that lava once extended across the present location of the river and covered the older rocks of the Cascade Range. The river subsequently cut its way through the basalt into the hard metamorphic rock below, forming today's steep-walled valley.

The Swakane Biotite Gneiss was initially a Cretaceous-age sandstone that experienced unusually high pressures and temperatures when the North Cascade microcontinent collided with North America. During metamorphism the original sand grains were recrystallized, and the minerals were aligned into light bands of quartz and feldspar and dark bands of biotite and amphibole. Roadcuts along this stretch of US 2 provide some excellent exposures of this beautifully layered banded gneiss. Quartz and granite dikes and sills have also intruded the rock, with sills parallel to the layers and dikes cutting across the layers.

Turtle Rock may be an in-place remnant of the gneiss bedrock, but most geologists think it is an unusually large landslide that slid down the steep eastern valley wall. Huge rotated blocks of bedrock on either end of the debris pile create a profile reminiscent of a giant turtle. Did the slide completely block the Columbia River? We don't know because the river, with the aid of a catastrophic flood that

Turtle Rock, along the Columbia River, is the remnant of a large rockslide.

A large ice age flood bar (in photo center) displaying giant current ripples (vertical lines) is reflected in the quiet waters of Lake Entiat across from the viewpoint on US 2. Floodwater here flowed from right to left.

Swakane Biotite Gneiss is exposed in roadcuts along US 2 between East Wenatchee and Turtle Rock.

roared down the valley during the late phases of the last glacial episode, removed most of the debris and deposited an elongated sand and gravel bar on the downstream end of the island. This last flood resulted from the catastrophic draining of Glacial Lake Columbia when the Okanogan ice dam failed. Lake Missoula had disappeared a few hundred years earlier, and its floods never reached here when the Okanogan ice lobe blocked the Columbia River.

Across the river from the parking area, a side canyon is partially filled with a flood bar deposit from an ice age flood. Rapidly flowing water over 500 feet deep created giant ripple marks on the flood bar surface. Only the catastrophic release of a huge volume of water can create the hydraulic conditions necessary to form giant ripple marks.

39. MT. RAINIER
A Living Volcano beneath Glaciers

Volcanism at the modern Mt. Rainier cone began about a half million years ago, when extensive flows of andesite and dacite lava and pyroclastic flows spread out like spokes on a wheel, filling the surrounding valleys. Much later these resistant lavas would become the tall ridges radiating from Mt. Rainier's cone. Erosion preferentially attacked the adjacent weaker rock, so the ridges became valleys and the

valleys became ridges. Later, less voluminous, shorter lava flows built up around the vent and were interlayered with ash and other fragmental materials blown out of the vent during eruptions. The layers built a steep volcanic edifice known as a stratovolcano.

Mt. Rainier currently stands 14,411 feet above sea level, the tallest peak in Washington and in the entire Cascade Range. As recently as 10,000 years ago its elevation may have exceeded 15,000 feet. Topographic evidence indicates that the former summit was removed in an eruption. The 5,700-year-old Osceola mudflow contains sufficient volume

The debris-covered snout of Nisqually Glacier on the south side of Mt. Rainier is a few hundred feet thinner than it was in the mid-1800s. The nonvegetated lateral moraine along the valley side marks the nineteenth-century height of the glacier.

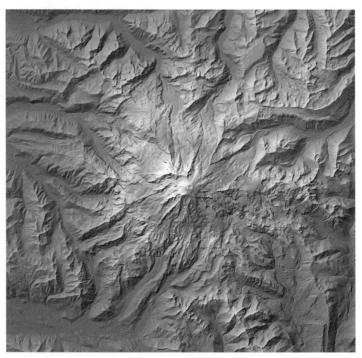

Ridges that radiate away from Mt. Rainier are former lava flows that filled valleys surrounding the mountain. Note the edge of the blown-off mountain top and the new summit cone forming in its center.

Remnants of the collapsed older cone (dashed lines) of Mt. Rainier occur as isolated rocky ridges on the mountain flanks.

to account for the missing top of the mountain. The mudflow reached Puget Sound some 70 miles away and covered an area where a few hundred thousand people live today. Following the summit collapse and mudflow, summit cone eruptions during the past 2,000 years built the upper 1,200 feet of the mountain into its current form. Nearly a dozen ash layers from local eruptions have been deposited around the mountain since the end of the ice age.

Evidence of potential activity includes one or more minor eruptions during the 1800s, the current emission of significant heat and gases from fumaroles in the summit crater, and the frequency of earthquakes directly below the cone. Earthquake activity at Mt. Rainier is second in frequency only to that at Mount St. Helens. The geologically recent events, the chemically weakened condition of some of Mt. Rainier's bedrock, the extreme steepness of its slopes, and its location close to a major population center make it one of the world's most potentially dangerous volcanoes.

The flanks of the mountain are clothed in a blanket of snow and ice, forming the largest single-peak glacier system in the contiguous United States. Tens of feet of snow accumulate on its flanks each winter. Until about the late 1800s more snow fell during the winter than melted during the summer. Very recent measurements by park geologists indicate that glaciers here are shrinking, as are most other glaciers around the world.

At the Palisades, beautiful columns developed in welded tuff as it cooled 650,000 years ago. —Photo by Barbara Kiver

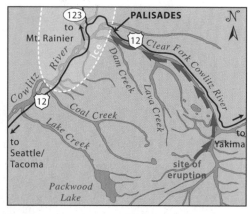

Dacite magma, erupted from the Goat Rocks Volcano, flowed down Clear Fork and pooled upon reaching a glacier in the Cowlitz drainage.

40. THE PALISADES
Columns in a Welded Tuff

Along the White Pass Scenic Byway (US 12) about 2 miles east of the WA 123 junction is an easy-to-miss rest area. A short entrance road leads to an outstanding view of the 650-foot-deep gorge of the Clear Fork Cowitz River eroded in columnar-jointed welded tuff. The tuff cooled from an unusually thick pyroclastic flow made of dacite, a volcanic rock relatively rich in silica. The dacite magma erupted as a very hot, gaseous flow that moved rapidly across the surrounding topography. When it came to rest the residual heat caused the ashy debris to initially melt before cooling into a solid body.

As a lava flow or ash flow cools and solidifies, it contracts, often producing a regular system of fractures that propagate through the cooling flow and tend to form four- to seven-sided geometric polygons, called columns. As the deposit loses heat to the underlying ground and to the atmosphere or cooler flow material above, cracks or joints form perpendicular to the cooling surface, resulting in vertical columns. The Palisades columns are unusually long and nearly geometrically perfect.

The ash flow erupted about 650,000 years ago from a vent about 9 miles to the east on the flanks of the now-extinct Goat Rocks Volcano. The unusual thickness of the flow here may be the result of ponding of the hot material in the lower Clear Fork drainage against the huge alpine glacier that existed in the Cowlitz River valley during the ice age. If this were indeed the case, the collision of hot ash and ice must have been a spectacular sight!

41. APE CAVE
Inside a Lava Tube

On the south flank of Mount St. Helens in the Gifford Pinchot National Forest, exceptionally fluid basalt magma erupted about 2,000 years ago. Lavas near the big volcanoes in the Cascade Range are usually rich in silica and mostly form andesite and dacite flows that lack the necessary fluidity to form a lava tube. However, this flank eruption produced basalt, and more than sixty lava caves or lava tubes formed in the Cave Basalt flow. At more than 2 miles long, Ape Cave is one of the longest continuous lava tubes in the world.

As flowing basalt cools, the moving lava quickly concentrates into a lava channel. If the lava continues flowing for a time, the solidifying edges of the flow will enlarge toward the center, forming a roof over the channel. The lava in the interior continues to drain downslope and can leave an evacuated section, or cave, behind. Different lava levels will form flow lines, or bathtub rings, on the cave wall. The last level is often marked by lava levees that resemble railroad tracks on the cave floor. The 1,000°F to 2,000°F heat inside the emptied flow often remelts the cave walls and ceiling, forming a glassy zone,

called a lava tube lining. While still hot, the sticky, plastic lining often sags and drips to form lava stalactites, or lavasicles, and lava stalagmites.

In the lower cave section, a lava boulder floating on a lava flow became wedged about 12 feet above the floor between walls that were cooling into a tube roof, forming the unique feature known as the Meatball. Lava tube ceilings can be quite thin, often collapsing to form a skylight, such as the one about halfway through Ape Cave. Although the interior of the cave was once a toasty location, the cave and surrounding rock materials have now taken on the seasonal average of 42°F. Warm clothing and three sources of dependable light are recommended for those who visit and explore this gem of a cave.

Ridges that look like bathtub rings on the walls of Ape Cave mark former levels of lava as the flow volume decreased inside the lava tube.
—Photo by John Scurlock

The nearly perfect cone of Mount St. Helens in 1972 prior to the 1980 eruption that blew off its top and obliterated its glaciers.

42. MOUNT ST. HELENS
New Kid on the Block

When Mount St. Helens burst into full eruption on May 18, 1980, it provided earth scientists firsthand experience of what can happen during a volcanic eruption in the Cascades. Scientists had already established that the volcano was very active and one of the youngest in the Cascades, no more than 40,000 years old. Stories passed down through the generations by Native Americans show they knew of the mountain's potential and gave it names such as Keeper of the Fire, Smoking Mountain, and Fire Mountain.

Mount St. Helens has experienced a number of eruptive intervals followed by dormant periods, some of which lasted for many thousands of years. By 1980 the nearly perfect cone stood some 9,677 feet above sea level. Numerous earthquakes beneath the mountain began on March 20, 1980, followed by the first steam eruption on March 27. Magma moved higher up into the mountain, creating a huge bulge on the north flank. On May 18, an earthquake shook the mountain, causing the bulge to collapse, taking the upper 1,300 feet of the summit with it in the world's largest historic rockslide. With the top removed, the underlying highly pressurized magma chamber below exploded to the surface, sending a blast wave laterally down the north flank as well as vertically. The blast bowled over and partly incinerated the north flank forest as much as 16 miles from the summit. Fine ash was blown into the upper atmosphere for nine straight hours, sending downwind areas into an eerie darkness. The gray ash settled to the ground, piling up as if it were snow. Some of the airborne ash lofted high enough that it eventually circled the globe.

Following the 1980 eruption, magma continued to erupt on a far less violent scale. The sticky, slow-moving magma formed a lava dome in the crater until 1986. All seemed quiet as earthquake activity reduced between 2000 and 2004 to its lowest level since the 1980 eruption. In September of 2004 a sudden increase in tremors and the formation of a bulge on the crater floor was followed a few days later by a series of explosions. Solidified but hot spines of lava punched up through the crater floor, building yet another lava dome before again returning to quiescence. The next chapter in Mount St. Helen's exciting history could begin at any time.

The post-eruption cone of Mount St. Helens, showing the collapsed summit and the steaming lava dome in the crater. —Photo courtesy of US Geological Survey

43. COLUMBIA RIVER GORGE
Here through the Ages

The Columbia River, the second largest river on the entire west coast of North and South America, has experienced many geologic upheavals. Volcanic eruptions, the formation of a mountain range in its path, massive ice age floods, and large landslides have all occurred in the 80-mile-long section known as the Columbia River Gorge, which cuts through the Cascade Range. The river has flowed through the general gorge location for at least 20 million years, and probably much longer. When Cascade volcanism began 40 million years ago, many rivers flowed westward across the young volcanic arc into the Pacific Ocean. Increasing elevations

in the Cascade Range eventually defeated most of the cross drainages until only the largest, the ancestral Columbia River, survived. Outpourings of the Columbia River Basalt from 17 to 6 million years ago completely changed drainage locations in the Columbia Basin to the east, but the Columbia River was able to persist at the gorge location despite being pushed around. Over twenty flows were large enough to flow the length of the ancestral Columbia valley, covering parts of what is now Vancouver, Washington, and Portland, Oregon. Basalt even reached about halfway down the Oregon Coast and flowed as far north as Grays Harbor

In the semiarid landscape of the east section of the gorge, the basalt flows dip gently to the east, on the limb of one of the Yakima Folds. Mt. Hood, a stratovolcano, is visible in the distance in Oregon.

View east from near Crown Point in Oregon. Beacon Rock, the neck of a very young volcano, juts up from the valley floor on the left and Bonneville Dam is visible upstream.
—Photo by Mark Kiver

along the Washington Coast. Although the river channel was buried by lava again and again, the river was trapped between its valley walls and cut anew into each of the flows with only minor changes in river location.

Volcanic activity occurred not only in the surrounding Cascade Range to the north and south but also in and near the Columbia Gorge. The formation of a volcanic field in the Portland area beginning 2.5 million years ago and Mt. Hood in the last 1 million years pushed the river northward. Eruption of the Beacon Rock volcano 60,000 to 50,000 years ago, the youngest of Portland's volcanic field, forced the river to flow around the cone to the south (discussed in site 44).

The massive Missoula Floods of the Pleistocene ice age were also funneled through the gorge, ripping loose volcanic and sedimentary material from its walls. The floodwater was over 1,000 feet deep in the gorge and flooded the future site of Portland under 400 feet of water. Significant landslides added some finishing touches to the landscape, including in the narrow river section where Bonneville Dam is now located (discussed in site 45). More recently, as development projects marred the relatively pristine landscape, Congress established the Columbia River Gorge National Scenic Area in 1986 to protect the quality of this special landscape.

Beacon Rock is the neck of a former volcano that erupted and diverted the Columbia River to the south. A very safe hiking trail leads to the top of this imposing monolith.

44. BEACON ROCK
Neck of a Former Volcano

Lewis and Clark and the Corps of Discovery camped within view of Beacon Rock on November 2, 1805. Although they named the rock, they were likely more excited about seeing for the first time that the Columbia River was responding to the influence of tides. The massive basaltic andesite volcanic neck towers 848 feet above the river and is all that remains of a former volcano that popped up in the middle of the Columbia River valley about 60,000 to 50,000 years ago, forcing the river to adjust its course. The volcano produced some lava flows and also considerable pyroclastic material. The core, or neck, of the volcano contained strong, solidified igneous rock. The less coherent material on its flanks was unable to withstand the onslaught of Missoula floodwater that roared through the canyon multiple times during the most recent glacial episode. The water velocity in the 1,000-foot-deep flood exceeded 70 miles per hour in places. Only the neck of the volcano remains today.

Henry Biddle purchased Beacon Rock in 1915 and built an amazing trail to its summit to enable the public to enjoy its spectacular views. Three years, fifty-three switchbacks, twenty-two bridges, and $10,000 went into the project. The fate of Beacon Rock was uncertain because as recently as 1931 a railroad and the Corps of Engineers considered it a potential source of rock for the railroad and also the jetty at the mouth of the Columbia River. In addition, a governor prevented Beacon Rock from becoming a state park until 1935, when a new administration was elected and the property was finally deeded to the state of Washington by Biddle's family.

45. CASCADE LANDSLIDES
Bridge of the Gods

A large landslide blocked the Columbia River about 250 to 550 years ago and created a 70-mile-long lake in the Columbia Gorge. Oral traditions of Native Americans tell of walking across the Columbia River on what they called the Bridge of the Gods. A forest was drowned behind the landslide, and when the lake overflowed the dam, an extremely violent rapid formed across its lower end. The Cascade Rapids, which now lie hidden beneath the reservoir behind Bonneville Dam, inspired early explorers to name the surrounding mountains the Cascade Range.

Although the Columbia River Basalt is a solid unit more prone to rock falls than landslides, it overlies less coherent 25- to 20-million-year-old volcanic sediments and mudflows erupted from nearby Cascade volcanoes. The southward tilt of these sedimentary rocks toward the Columbia Gorge contributes greatly to the instability of the canyon walls on the north side of the river. Saturation of the sediment with groundwater can cause some of the volcanic sediment to behave as a nearly frictionless liquid on which the solid rock above slides toward the river.

Four or more landslides make up the Cascade landslide complex. The jumbled landslide topography below Table Mountain and Red Bluffs is very distinct, and the scarp, the cliffs below the peaks, looks very young. Radiocarbon dates and tree-ring dates from trees buried by the landslide suggest that the most recent slide, the Bonneville landslide, occurred 250 to 550 years ago. Perhaps triggered by the jolt of an earthquake, the landslide flowed up to 3 miles from its source, burying the Columbia River channel. Minor downslope movement in the slide complex continues today.

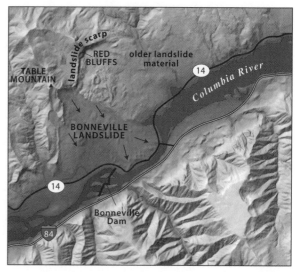

The Bonneville landslide pushed the Columbia River to the south side of the valley.

The mass of slide debris extends from the Columbia River near the foreground through the hummocky landslide topography in the middle distance to the scarp at Table Mountain (left peak) and Red Bluffs (right peak). Columbia River Basalt flows cap the valley walls here.

PUGET SOUND
Glacial Ice in a Tectonic Trough

During the past 15 million years as the Coast Range was dragged to the northeast by the subducting Juan de Fuca Plate, it was rotated in a clockwise direction. The shearing motion caused faults to form in the area between the Coast Range and the Cascade Range, and crustal blocks dropped down to form the Salish Sea lowland, including Puget Sound. As oceanic crust and marine sediment are stuffed beneath the edge of North America along the subduction zone, the Coast Range is being pushed up and the area to the east of the Coast Range is correspondingly depressed, with the faulted bedrock of the lowland as much as 3,000 feet below sea level. Sediment eroded from the surrounding hills, ash and volcanic sediment from nearby erupting volcanoes, and mud deposited on the seafloor has filled the fault basins with thousands of feet of sediment. Occasional earthquakes, especially in the Puget Sound area, indicate that those faults are still active. The damaging magnitude 7.1 earthquake in Olympia in 1949, the 6.5 quake in Seattle in 1965, and the 6.8 quake in Nisqually in 2001 are ominous reminders of what can and will happen in the near future. The Nisqually earthquake caused up to $4 billion of damage and injured about four hundred people.

During the ice age, the Puget ice lobe, a finger of the 1-mile-thick continental glacier, followed this tectonic lowland. The harder bedrock in the San Juan Islands and northward was scraped and sculpted by the flowing ice.

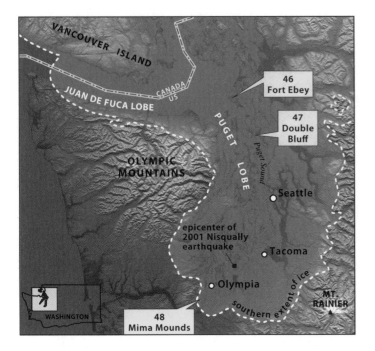

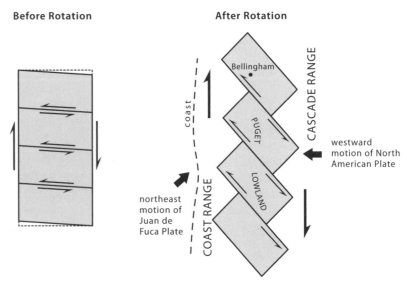

Oblique plate motion rotates crustal blocks along a series of faults in the lowland between the Coast Range and the Cascades. —Diagram modified from Alt and Hyndman, 1984

Farther south in Puget Sound, the younger, less resistant rock was eroded deeply by the advancing ice. As the Puget ice lobe eventually thinned, slowed, and became less erosive, it and its meltwater deposited glacial material, lake sediments, and outwash in the Puget Sound area. This loosely indurated sediment, which is widely exposed in the surrounding sea cliffs and along the edges of numerous islands, collapses easily with wave erosion, earthquakes, and tsunamis.

A large glacial erratic of Chuckanut sandstone on Double Bluff Beach on Whidbey Island was transported by ice about 50 miles from its bedrock source near Bellingham.

Barricades at Camano Island State Park prevent travelers from driving on a section of road that is slipping toward the sea. A steep sea cliff composed of poorly indurated Pleistocene sediments and clay layers is just out of view to the right.

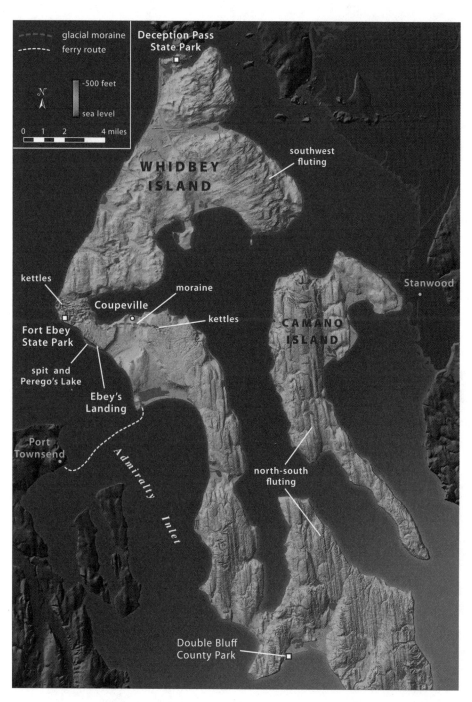

46. FORT EBEY
Ice Lobe Antics on Whidbey Island

Along the western shore of northern Whidbey Island, the Puget Sound lowland connects with the Strait of Juan de Fuca, the deepwater passage to the open Pacific Ocean. During the last glacier advance about 18,000 to 15,000 years ago, ice originating in the Coast Range and the Fraser River valley in British Columbia moved south and split into two lobes: the Juan de Fuca lobe flowing westward out the Strait of Juan de Fuca, and the Puget lobe continuing south to where it terminated near the present location of Olympia.

We know the flow directions because the ice formed flutes, elongated south-trending topographic hills separated by swales, as it streamlined the surface of Whidbey Island. The Puget ice lobe was extremely heavy, creating enormous pressure on the land beneath it. At Fort Ebey the ice was about 4,000 feet thick, pushing down with a force of about 1,300 tons per square meter. Because the island is either forested or altered by agriculture, the fluting is best seen on topographic maps and LIDAR images, a special type of image that portrays the land surface without the complications of vegetation or buildings.

This LIDAR image of Whidbey and Camano Islands shows the prevalence of north-south ridges, linear topographic fluting produced by the override of a thick glacier. Note the area where a block of ice stagnated and melted, forming hundreds of kettles.
— *LIDAR image courtesy of Washington Division of Geology and Earth Resources*

Labels on map: glacial moraine; ferry route; –500 feet; sea level; 0 1 2 4 miles; Deception Pass State Park; WHIDBEY ISLAND; southwest fluting; kettles; moraine; Coupeville; kettles; Stanwood; CAMANO ISLAND; Fort Ebey State Park; spit and Perego's Lake; Ebey's Landing; Port Townsend; Admiralty Inlet; north-south fluting; Double Bluff County Park

During glacier retreat, global sea level rose significantly, causing the Juan de Fuca ice lobe to become a rapidly retreating tidewater glacier, resembling those in Glacier Bay, Alaska, today. With the elimination of the Juan de Fuca lobe about 16,000 years ago, the remaining body of the Puget lobe changed direction and re-advanced rapidly to the southwest, creating a series of southwest-oriented hills separated by swales on the northern part of Whidbey Island. As the glacier thinned and was unable to sustain the new flow, it deposited a moraine in the Coupeville area and left a large body of detached ice in the Fort Ebey area. Sediment from the melting glacier buried the stagnated ice mass. As the buried ice melted, sediment slumped, leaving behind a maze of hundreds of basins, or kettle holes. Trails wind around and across the forested kettles in Fort Ebey State Park and the adjacent Kettles Trails county park.

Waves and currents of the rising sea attacked the new landscape, forming impressive bluffs and beaches at Fort Ebey. Longshore currents moved sediment eroded from the bluffs southward toward Ebey's Landing National Historical Reserve, where a finger of sand, called a spit, formed. The spit grew outward and eventually reattached itself to the main island to form Perego's Lake, a water body isolated from the saltwater of Admiralty Inlet. Washovers during storms create a brackish environment in the lagoon.

View of the south end of Peregos Lake from the Bluff Trail at Ebey's Landing National Historical Reserve.

Waves eroding the base of the weakly consolidated sediments, along with numerous springs and seeps, all contribute to frequent slope failures at Double Bluff. Note the dark peat layer in the upper left corner.

47. DOUBLE BLUFF
Comings and Goings of Ice Sheets

The most recent glacier that occupied the Puget Sound area, about 18,000 to 15,000 years ago, mostly destroyed evidence of previous glacial and interglacial events, but remnants are present at Double Bluff County Park along the southwest edge of Whidbey Island. A spectacular sea cliff with a stack of sediments over 300 feet thick provides a glimpse of several older glacial events. Like a book resting on its side with page one on the bottom, the sediment layers, or pages, reveal the story of the past 200,000 years. Unfortunately, some of the pages here are missing because of erosion. The full story is patched together from evidence gathered at different localities in the Salish Sea area.

The oldest sediment at Double Bluff, an unsorted accumulation of clay, silt, pebbles, and boulders known as till, was deposited about 200,000 years ago by a glacial advance known as the Double Bluff glaciation. The glacial till is exposed at the bluff point on the west end of the beach, some 2 miles from the parking lot. Overlying these glacial sediments is the 120,000-year-old, interglacial Whidbey Formation, which includes stream, lake, and swamp deposits, now beds of sand, silt, and peat, a low-grade coal. As the cliffs erode back, pieces of very dense peat, some over 20 feet long, have fallen down and now lie scattered about on the beach. The former swamp deposit was converted to peat by the weight of over 100 feet of sediment plus two later ice advances in which glaciers over 3,000 feet thick added another 100 tons of weight per square foot. Overlying these are delta sands and silts deposited by mainland rivers into a freshwater lake and later into the ocean as sea level rose with the melting of the glaciers. Erosion of the cliff by waves continues to expose fresh sediments, and storm waves and longshore currents continue to move loosened sediment from the cliff base, forming an outstanding beach. At the present rate of erosion, the cliff retreats backward 100 to 200 feet every century.

Highly contorted folds in the sediment layers, visible at the base of the bluff, formed shortly after the sediments were deposited. The water-saturated sediments were violently shaken by a major earthquake on a nearby fault. The sediments were originally deposited in nearly horizontal layers, but now they are distorted and fractured into chaotic patterns. The earthquake may have occurred on the nearby South Whidbey fault, which shows evidence of recent movement.

Large pieces of peat tumbled down onto the beach at Double Bluff. Plant fragments and woody debris accumulated in swamps and marshes about 120,000 years ago during a warm interval between glaciations and were later compressed into peat, a low-grade coal.

An earthquake likely shook these beds shortly after deposition, contorting a 15-foot-thick zone at Double Bluff.

93

48. MIMA MOUNDS
Enigmatic Hummocks

Mima Mounds Natural Area Preserve, south of Olympia in the Chehalis–Black River drainage basin, preserves some of the original prairie in the Puget Sound lowland much as it appeared when the first explorers arrived. The Mima Prairie surface is covered with hummocks, or mounds, that are a few feet high and up to a few tens of feet in diameter. Settlers in the Puget Sound area flattened an untold number of these once-abundant mounds to better adapt the land to agriculture.

The southern edge of the Puget Lobe glacier was a short distance south by Olympia about 16,000 years ago. As the glacier front retreated northward, outwash streams deposited the sand and gravel that form the Mima Prairie surface, but how they ended up in mounds is not definitely known. Did water erosion remove the sediment between the mounds? Did a freeze-thaw process or some other cold-climate process cause the mounds to form? Did some gopher-like animals build these gravel mounds for homes and lookouts to spot predators? Did an earthquake shake the loose sediment, causing it to rearrange into hummocks? Or did certain plants or groups of plants preferentially protect the soil while areas around them were eroded away? We still don't know. The formation of Mima mounds has perplexed geologists and others since their first description in 1792 during George Vancouver's exploration of the area. Similar-appearing mounds are found in diverse localities around the world. Mounds in the Columbia Plateau on the east side of the state are composed of silt with no gravel, and wind likely played a major role in their formation. Wind would have a hard time moving the pebble-size material in the Mima mounds here, however. Trails wind through the natural area, and an informative kiosk offers possible explanations. To reach the area, take exit 95 from I-5 and head west through Littlerock to a T intersection. Turn right on Waddell Creek Road and look for signs for a left turn in about 1 mile.

Mima Prairie mounds. —Photo by Barbara Kiver

SAN JUAN ISLANDS
Stacked Thrust Sheets Scraped by Ice

About 172 islands form the San Juan Islands, with another 250 rocky knobs protruding above the sea during low tide. The islands are all mountaintops now partially flooded by the sea. The San Juan Islands National Monument, established in 2013, protects many rocky islands and points scattered throughout the island group. Most of these areas were originally reserved for navigation aids, and many still serve that function. The Patos and Stuart Islands units require private boat access, but units on San Juan and Lopez Islands can be reached easily by car via Washington State Ferry. In this chapter we also cover a few sites on and near Fidalgo Island, which is not technically one of the San Juan Islands but shares their geology.

The island group is a geologic mystery that only began to make sense after scientists' understanding of plate tectonics improved and countless hours of fieldwork were invested. Wayward masses of crustal rocks, called terranes, moved eastward about 100 to 65 million years ago and were shoved against and beneath the continent edge along a subduction zone. Some terranes were piled up in the western North Cascades along faults, and a stack of flat-lying thrust sheets—each a separate terrane—extends from the western part of the North Cascades across the San Juan Islands. In profile the thrust sheets in the islands resemble a fallen stack of books whose surfaces are slanting gently to the east. Each sheet is separated from the next by a thrust fault that slid one slice of rock over another. If the horizontal movement of each fault slice in the San Juan Islands and the western

This pillow basalt, exposed in the Watmough Bay unit of the San Juan Islands National Monument on Lopez Island, formed during a seafloor eruption along a spreading ridge or submarine volcano.

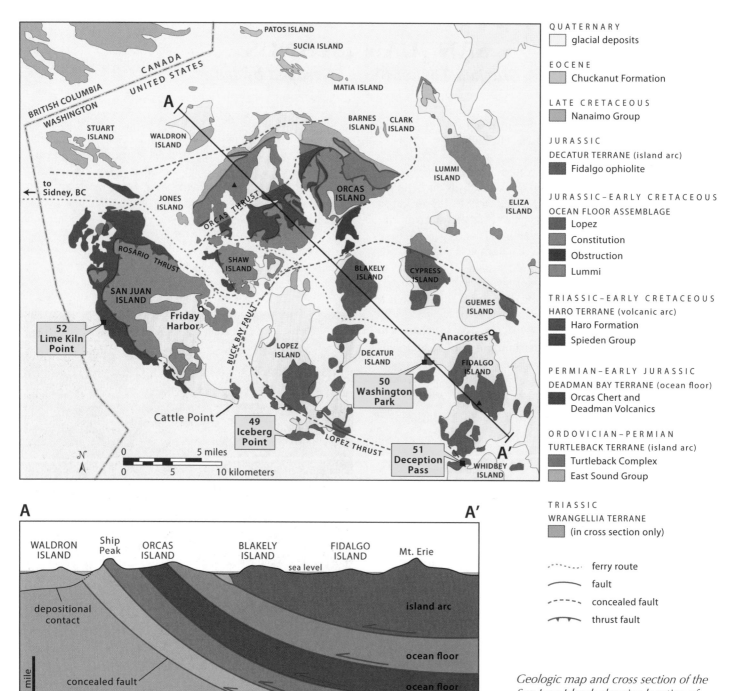

QUATERNARY
glacial deposits

EOCENE
Chuckanut Formation

LATE CRETACEOUS
Nanaimo Group

JURASSIC
DECATUR TERRANE (island arc)
Fidalgo ophiolite

JURASSIC–EARLY CRETACEOUS
OCEAN FLOOR ASSEMBLAGE
Lopez
Constitution
Obstruction
Lummi

TRIASSIC–EARLY CRETACEOUS
HARO TERRANE (volcanic arc)
Haro Formation
Spieden Group

PERMIAN–EARLY JURASSIC
DEADMAN BAY TERRANE (ocean floor)
Orcas Chert and Deadman Volcanics

ORDOVICIAN–PERMIAN
TURTLEBACK TERRANE (island arc)
Turtleback Complex
East Sound Group

TRIASSIC
WRANGELLIA TERRANE
(in cross section only)

········ ferry route
———— fault
– – – – concealed fault
⌢⌢⌢ thrust fault

Geologic map and cross section of the San Juan Islands showing location of major thrust faults (teeth are on the upper plate). —Modified from Brown, 2014

North Cascades were added together, it is estimated that mountain building shortened the Earth's crust here by at least 200 miles.

None of the rocks in the thrust sheets are homegrown; all formed somewhere else, either pieces of land sliced off a continent or volcanic islands or sections of ocean floor that were transferred a thousand or more miles to the west coast of North America. Although the rocks were metamorphosed and deformed in the mountain building, fossils that can still be identified tell of creatures that lived in warmer climates closer to the equator.

Once assembled on the west coast of the growing continent, sediments were deposited along the shoreline in Cretaceous time. Sandstone from the time when dinosaurs still lived can be found in a few places on the San Juan Islands and adjacent coast.

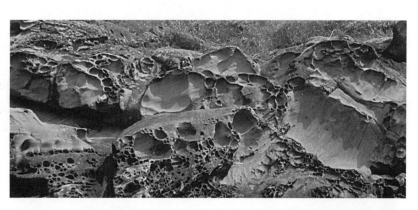

The distinctive weathering characteristics of the Chuckanut sandstone of Cretaceous age, shown here on Patos Island, are attributed to saltwater erosion.

These fractured and contorted layers of interbedded chert and sandstone of the Deadman Bay terrane are located along the Lopez thrust fault zone just east of Iceberg Point. Chert is common in the San Juan Islands and forms from deepwater deposits. The tiny silica shells of plankton rain down on the deep ocean floor, forming layers of silica ooze, which eventually harden into chert, a type of quartz with extremely tiny crystals. These rocks were brought to the surface during the subduction and thrust fault process.

Glacial groove cut into sandstone-mudstone rock at Iceberg Point on Lopez Island.

One of the large rocks that gouged this glacial groove at Iceberg Point rests in the groove next to the man.

49. ICEBERG POINT
Grooves and Erratics of Ice Long Gone

Iceberg Point, a unit of the San Juan Islands National Monument on Lopez Island, is an excellent place to see glacially scoured rock of the Lopez terrane. After forming about 150 million years ago in the deep ocean, the sandstone, chert, mudstones, and seafloor basalts were carried into the subduction zone, scraped off about 100 to 65 million years ago, and thrust westward. Iceberg Point is near the edge of the Lopez thrust fault, and the sedimentary rocks dip steeply into the island. The fault runs along the west edge of the island and extends northward between Lopez and San Juan Islands. The north tip of Cattle Point on San Juan Island is also part of the Lopez terrane.

Millions of years after the rock was added to the western edge of North America, the continental ice sheet of Pleistocene time polished and grooved the bedrock and deposited till and glacial erratics, rocks carried by the ice from somewhere along the glacier's path. Light-colored granitic boulders scattered about the landscape were carried by the glacier from the Coast Range of Canada, a minimum of 200 miles of travel distance. They were encased in the ice and then left in place as the ice melted. Some of the rocks encased in the base of the ice scratched and polished the bedrock, even scouring deep grooves, several of which can be found at Iceberg Point.

View east over part of the American Camp unit of San Juan Island National Historical Park to the Cattle Point unit of San Juan Islands National Monument on the south end of San Juan Island. The Pleistocene-age glacial sediment here was deposited directly on the dark, 150-million-year-old sandstone of the Lopez terrane, visible along the water's edge.

Granitic glacial erratics (the large whitish rocks) were transported to Iceberg Point from the Coast Range in British Columbia.

50. WASHINGTON PARK
View into the Mantle

On Fidalgo Island in the city of Anacortes is a 220-acre wooded peninsula set aside as Washington Park. Along the west side of the park is a rarely exposed ophiolite, the sequence of rock that makes up oceanic crust, including part of the underlying mantle. We don't often get a chance to see the deeper parts of ocean crust because they're 3 to 5 miles below the ocean floor and usually get consumed in subduction zones. Denser oceanic crust normally slides beneath the lighter, more buoyant continental crust. Here, a volcanic seamount along with some of the mantle was sheared off during the subduction process and later incorporated into the Decatur terrane, which was mashed into North America about 90 million years ago.

The complete ophiolite sequence ranges from peridotite, the dense mantle rock; up through basalt and gabbro, emplaced along a spreading ridge or oceanic volcano; to the cap of shale and chert, deep-sea sediments that accumulate on the newly formed oceanic crust. The entire ophiolite sequence is exposed on Fidalgo Island, with the rarest part, the mantle rock, visible in Washington Park, particularly at West Beach and Green Point along the north shore. The peridotite, which is partially metamorphosed to serpentinite here, is rich in the heavy elements iron and magnesium, with traces of nickel and cobalt. In a fresh exposure the rock is dark green to black, but it becomes buff colored when weathered.

Glacially scratched and polished rock surfaces are well exposed at West Beach where storm waves have stripped away glacial till that once covered the serpentinite. When the fresh-appearing scratch marks or striations are exposed, they weather rapidly and in a few years become less distinct and disappear. An unusually large glacial groove on the south side of the park was formed about 17,000 years ago, when a large boulder or group of rocks embedded in the base of the 1-mile-thick glacier scraped across the serpentinite surface.

Look for this remarkable glacial groove with fine scratch marks along one of the trails on the south side of Washington Park. An overhanging serpentinite roof protects the groove from rapid weathering.

Mantle rock exposed at low tide on West Beach at Washington Park has been altered to the rock serpentinite. The yellow bands are entirely the mineral serpentine, and dark layers are a mix of serpentine and a dark mineral called pyroxene. Note the 3-inch lens cap for scale in lower left of close-up.

The Deception Pass Bridge was built in 1935 and connects Fidalgo Island on the left with Whidbey Island on the right.

51. DECEPTION PASS
A Terrane Boundary

Deception Pass, a very narrow strait between Whidbey and Fidalgo Islands, was named by Captain George Vancouver in 1792 after members of his crew missed it the first time by. Deception Pass State Park, which spans the strait, is a good place to see bedrock, rocky shorelines, and small pocket beaches and sandbars. Compared with the extensive beaches and high bluffs of Pleistocene-age sediments to the south, there are only thin glacial deposits here.

Highly fractured and sheared basalt and ocean floor sediments like chert, shale, and a type of sandstone called graywacke make up most of the bedrock in the park. Individual rock layers cannot be followed very far before a

fracture or fault is encountered. The rocks look as if they were squeezed in the jaws of a vice, and indeed, they were. As the eastward-moving oceanic plate reached the subduction zone, oceanic crust, submarine volcanoes, and the overlying marine sediments were scraped off and shoved eastward, colliding with other scraped-off rocks, creating a mess like a giant freeway accident. A series of thrust faults formed, creating the separate terranes that make up the San Juan Islands. Two of those terranes, the Decatur and Lopez, join along the seaward edge of Deception Pass State Park. A major fault likely extends through Deception Pass, and erosion of broken rock along the fault helped form

A low-angle fault east of Lighthouse Island along Deception Pass is part of the thrust fault zone that separates the Lopez and Decatur terranes. The fault slants down from the upper left to lower right. Note the sea cave that formed in the fractured and weakened rock along the fault.

this canyonlike gap between Fidalgo and Whidbey Islands. The huge Puget ice lobe undoubtedly played a major role in the canyon excavation.

Waves and currents have put the finishing touches on the present-day shoreline, producing steep, rocky cliffs broken by small pocket beaches and sandbars. West Beach at Deception Pass is a wide accretionary beach, one that has grown seaward with time as longshore currents bring sediment north from eroding bluffs on Whidbey Island. An attached spit has blocked off an inlet that now contains Cranberry Lake. South of the lake along West Beach, sand behind the present beach is blown up into a number of dune ridges that extend nearly a half mile to the south of Cranberry Lake. The ridges are mostly stabilized by vegetation, including invasive European dune grass, but some free-moving sand is present. One of the stabilized ridges on the back beach has a Douglas fir that is more than 850 years old, indicating that the beach and dune field have likely been here for 1,000 years or more.

This Douglas fir tree in the stabilized back dune area south of Cranberry Lake is over 850 years old. —Photo by Barbara Kiver

52. LIME KILN POINT
Out-of-Place Limestone

Pods and layers of high-quality limestone on San Juan Island were a major source of cement during early settlement days in the Pacific Northwest. Quarrying and construction of limekilns began in 1860 at Lime Kiln and nearby Roche Harbor on San Juan Island. The 1906 San Francisco earthquake fueled the quarrying, as demand for cement increased dramatically along the Pacific Coast. The kilns closed at Lime Kiln in 1920, but quarrying continued until 1950.

Most of the rocks in the San Juan Islands originated in the deep ocean, where limestone does not form. Invertebrate shells and shell fragments made of calcium carbonate, or calcite, make up most limestone deposits. Calcite dissolves in cold acidic water about 12,000 feet or more below the surface, so marine limestone only forms in shallow-water environments. Even in the middle of an ocean, however, submarine volcanoes like those in Hawaii reach shallow depths and provide habitat where lime-secreting organisms flourish and limestone can form. The invertebrate fossils in the limestone at Lime Kiln are about 280 to 260 million years old and are similar to those found in equatorial regions in the western paleo–Pacific Ocean, suggesting that the rocks here were transported eastward and northward several thousand miles before becoming part of the growing North American continent.

Restored limekiln at Lime Kiln Point State Park.

A pod of light-colored limestone is exposed on the west side of San Juan Island amidst volcanic rock. Lime Kiln Lighthouse, in the distance, was constructed in 1919.

OLYMPIC MOUNTAINS
A Youthful Coast Range

"Geologizing," a term used by Charles Darwin, is not easy in the Olympic Mountains, where intense glaciation by both mountain and continental glaciers left a cover of loose glacier- and stream-derived sediment, obscuring the underlying rock. The thick cover of vegetation in the Olympic rain forest, which receives about 150 inches of precipitation per year, also greatly impedes geologic studies. However, good exposures of bedrock occur near and above timberline, where heavy snow accumulations help maintain the lowest-elevation glaciers in the contiguous United States. Along the coast, the fury of the Pacific Ocean has eroded a wonderful variety of rugged coastal landforms, including sea cliffs, headlands, wave-cut benches, sea stacks, sea arches, and sea caves, many within Olympic National Park.

Jagged peaks soar to 7,980 feet in the Olympic Mountains, which have been uplifted in only the last 15 to 10 million years and are still rising. A thick section of ocean-floor sediments has been stuffed into the subduction zone and uplifted, becoming the crumpled and tortured rocks in the Olympic Mountains. The ocean floor sediments here are being pushed beneath a slab of basaltic oceanic crust called the Crescent Formation, lifting it into a dome. Erosion has

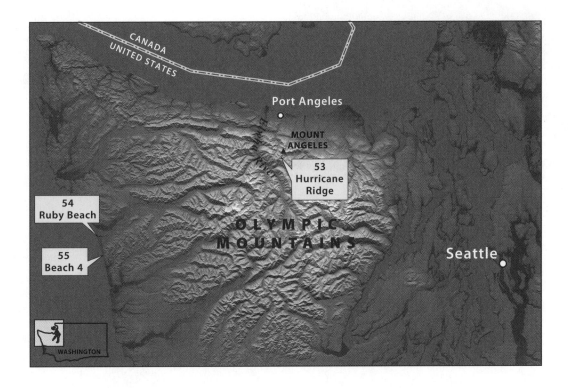

View into the mountain interior from Hurricane Ridge in Olympic National Park.

removed the top of the dome leaving the Crescent Formation exposed in a horseshoe-shaped outcrop pattern that opens toward the sea.

The geologically youthful streams draining the Olympic Mountains are important spawning grounds for salmon. The Elwha River, which flows north to the Strait of Juan de Fuca, hosted two dams until 2012. The National Park Service removed the illegally constructed Elwha Dam, built in 1913, and Glines Canyon Dam, built in 1927, to restore the river and its salmon runs. Over 70 miles of prime salmon spawning streams capable of producing almost 500,000 fish per year are now available. For the century the dams were in place, sediment accumulated on the reservoir bottoms. Now the river is removing the sediment, which is replenishing beaches along the strait.

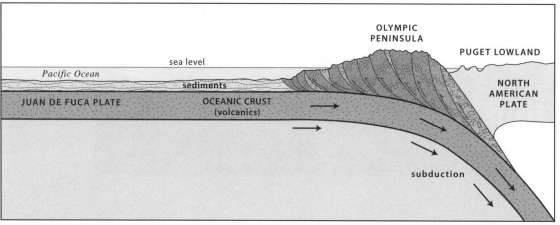

Marine sediments are being scraped off along the subduction zone, rotated upward to near vertical positions, and elevated to nearly 8,000 feet in the Olympic Mountains. —Modified from Rau, 1980; Alt and Hyndman, 1995; Brown, 2014

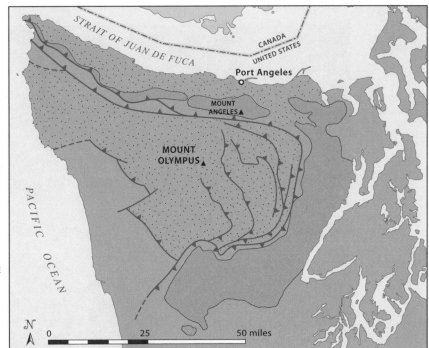

Generalized geologic map of the Olympic Peninsula. — Modified from Alt and Hyndman, 1995

EOCENE
Crescent Formation

EOCENE to OLIGOCENE
sedimentary rocks

——— fault
- - - - - concealed fault
⊾⊾⊾⊾ thrust fault

0   25   50 miles

The drained lakebed behind the former Elwha Dam as it appeared in 2012, months after the dam was removed. In the area's wet environment it won't take vegetation long to reclaim the barren lake bottom.

Steeply dipping sedimentary rock deposited on the deep ocean floor now forms the summit of Mt. Angeles in Olympic National Park, over 6,000 feet above sea level. —Photo by Bill Baccus, National Park Service

53. HURRICANE RIDGE AND MT. ANGELES
Tilted Rock from the Ocean Floor

Slices of Eocene-age oceanic sediments, rotated to nearly vertical or overturned positions, can be found in the core of the Olympic Mountains. A 17-mile-long road climbs from the Olympic National Park Visitor Center in Port Angeles to the top of Hurricane Ridge, where short hikes to Hurricane Hill and Mt. Angeles provide access to the twisted rock wreckage. The stunning peaks and glaciers in the core of the Olympic Mountains are spread out in plain view, and broad, deep valleys formerly occupied by ice age glaciers lie below.

Sand and mud deposited on the ocean floor were carried, conveyor-belt-style, on the basaltic ocean plate moving eastward into the subduction zone. The sediments were changed to rock under the weight of thousands of feet of overlying sediment, compressive tectonic forces, and secondary mineral cementation. Sediments that were scraped from the moving oceanic plate were stuffed under the continental plate, thickening the crust and tilting layers on end. Slowing of plate movement about 12 million years ago allowed the thickened crust to float higher and become towering mountains.

54. RUBY BEACH
Mélange of Sea Stacks

Ruby Beach, one of the most scenic beaches on the Olympic coast and part of Olympic National Park, is reached by a short trail from US 101. The rocks here are part of the Hoh Rock Assemblage, a group of marine sedimentary and volcanic rocks that are highly fractured and mangled by the tectonic forces that built the Olympic Mountains. The numerous isolated sea stacks scattered across the beach and tidal area are huge chunks of volcanic rock, entirely different from the fractured sandstone along the sea cliff behind the beach. Geologists suspect that this section of the Hoh Rock Assemblage is a tectonic mélange, a mixture of large, intact blocks of rocks surrounded by weak siltstone or mudstone. Imagine the forces required to fracture and mix rocks as if the large rocks were peanuts in chunky peanut butter. Removal of the weaker rock material frees the large chunks—the sea stacks—some of which are bigger than houses.

Bedrock behind the beach has been planed off into a nearly level surface. Nearby Abbey Island has a flat top at the same elevation. These surfaces are called wave-cut platforms and formed when the land surface was at sea level and waves beveled the bedrock. The flat surface is now above sea level as mountain-building forces continue to elevate the land.

Ruby Beach in Olympic National Park has numerous sea stacks and wide beach areas that are readily accessible during low tide. The flat top of Abbey Island (left side) is a wave-cut bench that was cut at sea level by wave action and then uplifted.

55. BEACH 4
Upside-Down Sandstone

At Beach 4, south of Ruby Beach, well-bedded sandstones and siltstones are more coherent than those of the tectonic mélange found at Ruby Beach about 4 miles to the north. Despite their intact layers, however, the rocks are upside down. Sediment-laden turbidity currents, basically submarine landslides, deposited the sand and mud in horizontal layers on the floor of a deep sea about 15 million years ago. We know the rocks are overturned because sedimentary features, such as graded bedding, are upside down. When submarine landslides race down the continental slope into deeper water, the large particles fall first from the turbidity current. As the sediment cloud continues to settle, increasingly finer particles follow, forming the graded bed. Because the beds here are overturned, the finer particles that mark the tops of the beds are overlain by the coarser particles.

The inclined beds at Beach 4 were truncated by waves when the rocks were at sea level. The flat surface, known as a wave-cut platform, is now about 10 to 20 feet above high tide. Elsewhere along the Olympic coast wave-cut platforms range from below sea level to as high as 160 feet above high tide level, indicating substantial differences in coastal uplift during the past few hundred thousand years. Flat-lying glacial deposits cover the tilted rocks along the wave-cut platform in places, forming an angular unconformity. At least 15 million years have passed between the time the layered sedimentary rock was deposited and when streams fed by glaciers deposited the flat-lying sediment over the now tilted and planed-off rock.

Close-up of overturned sandstone and siltstone beds on Beach 4 in Olympic National Park. Each sand and shale couplet was deposited during a single submarine landslide. Note 3-inch lens cap for scale.

Close-up view of holes bored by piddock clams when the sandstone was in or below the tidal zone. Recent tectonic uplift has elevated the rock above high tide level, where intertidal animals cannot survive.

The tilted edges of sandstone beds of Miocene age were planed off by wave erosion, uplifted, and then covered by overlying, horizontal glacial sediments deposited during Pleistocene time.
—Photo by Barbara Kiver

Movement along this low-angle fault caused rock layers to crumple into tight folds on Beach 4 in Olympic National Park.
—Photo by Barbara Kiver

FAULT

WILLAPA HILLS
Sands of the Southwestern Coast

The Coast Range south of the Olympic Mountains is a much lower area called the Willapa Hills, with elevations only up to 3,110 feet. Like the Olympic Mountains to the north, the Willapa Hills are also dominated by Eocene volcanic and sedimentary rocks that formed on the ocean floor and have subsequently been tectonically uplifted. However, the Willapa Hills do not display the intense folding and metamorphism that is seen in the Olympic Mountains, although more intense deformation may be present at depth. The Willapa Hills and the coastal strip are not rising as rapidly as the Olympic Mountains and were partly drowned when sea level rose over 300 feet with the melting of Pleistocene

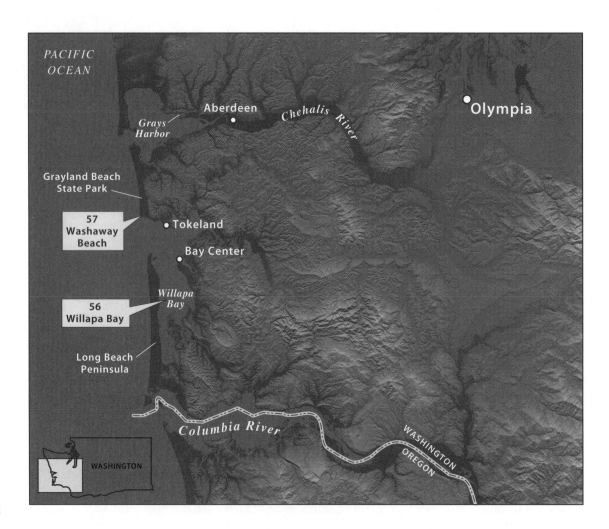

Wide beach at Grayland Beach State Park at low tide. Note the vegetation stabilizing the small dunes on the back beach and the heavy surf in the distance. —Photo by Barbara Kiver

View looking southeast from Tokeland near the mouth of Willapa Bay toward the Willapa Hills, the Coast Range here, in the distance.

View from near North Head Lighthouse looking northward at the Long Beach Peninsula, a very long spit that partially encloses Willapa Bay. The shallow sand shelf on the seaward side forms a wide zone of breakers. The spit protects the estuary on the far, lagoon side (out of view in this photo) from the heavy surf on the Pacific Ocean side. —Photo by Barbara Kiver

glaciers about 10,000 years ago. Long spits and sandy beaches contrast sharply with the rugged rocky coast along the Olympic Mountains.

Large volumes of sand carried downstream by the Columbia River form a large sandbar at its mouth, known as the Graveyard of the Pacific, where over two thousand ships have met their end since 1792. The dominant southwest winds along the coast here drive longshore currents to the north, which move sand. Much of the sand has been deposited along the 24-mile-long Long Beach spit that partially encloses the Willapa Bay estuary, the second largest and least disturbed estuary on the Pacific Coast. A 12-mile-long sand spit partially encloses the Grays Harbor estuary farther to the north. The sandy coast, with its wide beaches, stabilized sand dunes, and deposits from back dune swamps and marshes, has been growing for thousands of years, but conditions have changed over the past century. Beach environments are extremely dynamic and sensitive to small environmental changes, and these beaches are now eroding because of lack of incoming sand, possibly because of dams on the Columbia River and a jetty at its mouth.

56. WILLAPA BAY
Earthquakes, Tsunamis, and Ghost Forests

Offshore of the Pacific Northwest is the Cascadia subduction zone, the plate boundary along which the Juan de Fuca Plate is being thrust beneath the North American Plate. Geological evidence for extremely large earthquakes of magnitude 8.0 or above along the Cascadia subduction zone is recognized from British Columbia to northern California. Some of the best evidence occurs in Willapa Bay and Grays Harbor, where in the late 1980s Brian Atwater, Eileen Hemphill-Haley, and colleagues researched forest soils that had been abruptly covered by sand. This sudden burial occurred when the coast rapidly subsided during an earthquake and a tsunami wave deposited sand. Forests inundated by the saltwater quickly died, but tree stumps remain, leaving a record in their growth rings of exactly when some of these devastating events occurred. Fire hearths of prehistoric Native Americans have been discovered directly buried by sandy tsunami deposits along the Salmon River estuary in Oregon and the Niawiakum River in Willapa Bay, suggesting that there were human witnesses

to these natural disasters. Researchers have documented this kind of rapid burial of land by tsunami sands during two of the largest historic earthquakes, the 1960 Valdivia earthquake and tsunami in Chile (magnitude 9.5) and the Good Friday, or Great Alaskan, earthquake and tsunami of March 27, 1964 (magnitude 9.2).

Atwater's and others' research has found that multiple great earthquakes have shaken the Pacific Northwest every few hundred years for at least the past few thousand years. The most recent great earthquake was on January 26, 1700, based on the well-kept tsunami record in Japan and tree-ring studies along the Pacific Northwest coast. More than three hundred years have past since then, so it's very likely that another earthquake will shake the area again. If you are along the coast of Washington, Oregon, or northern California, you will notice "Tsunami Hazard Zone" signs in low-lying areas. If the ground shakes while you are near the ocean (or any large body of water) immediately head to higher ground. Tsunamis can be 100 feet tall and in deep

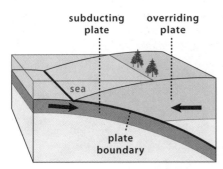 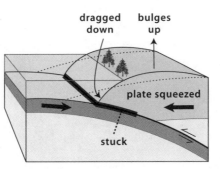 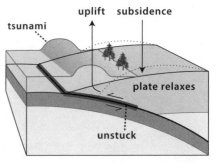

Extremely large earthquakes, known as great earthquakes, are associated with subduction zones in which the lower plate locks against the overriding upper plate, and the stuck, overriding plate bulges upward. When the plates slip, the stored energy along the fault is released in the form of seismic waves, causing an earthquake, and the bulged-up plate relaxes and subsides. How much fault slippage occurs determines the earthquake magnitude. The slip also may displace water in the ocean, causing a tsunami, which is as much of a hazard, if not more, than the earthquake itself. Because the land subsides, low-lying areas that were above high tide before the quake may be below sea level following the quake. —Modified from Brian Atwater, Orphan Tsunami, 2005

water move at speeds of over 500 miles per hour. People on the coast during an earthquake along the Cascadia subduction zone will have less than twenty minutes to evacuate low-lying areas.

Near the town of Bay Center in Willapa Bay are two areas in the tidal zone with tree snags sticking up—what geologists call "ghost forests." One area is reached by a trail in the Bush Pioneer Park campground in Bay Center, and the other is visible 1.4 miles south of the Bay Center bridge near Sandy Point. When the land subsided, probably during the 1700 AD earthquake, seawater inundated the forest, killing the trees. Overlying the ghost forest is a layer of sand deposited by the tsunami waves.

This ghost forest, south of Bay Center near Sandy Point in Willapa Bay, consists of mostly spruce trees killed during the Cascadia subduction zone earthquake on January 26, 1700.

Five buried soils (at arrows) are exposed at low tide at Johns River in Grays Harbor. Ages of the soils range from 1400 BC to 1700 AD. Hundreds of years separate each great earthquake. Note the buried logs projecting from some of the buried soils. —Photo courtesy of Brian Atwater, *Orphan Tsunami, 2005*

Remnant, spreading tree roots of a ghost forest in Bush Pioneer Park in Bay Center are visible at low tide.

57. WASHAWAY BEACH
There Goes the Backyard!

The aptly named Washaway Beach, on Cape Shoalwater north of the entrance to Willapa Bay, loses an average of 124 feet of width per year, the most rapid erosion on the US Pacific Coast. Sea cliff retreat has destroyed over fifty homes, along with part of the Willapa National Wildlife Refuge and a lighthouse, Coast Guard station, school, and cannery building. State Highway 105 had to be rerouted inland, and a historic cemetery had to be physically moved.

Rapid erosion began in the early 1900s and has continued for over a century. In the 1800s Cape Shoalwater projected out more than 2 miles from the present shore. Currently, much of the sediment moved north along the Long Beach Peninsula is carried offshore by currents or is washed into Willapa Bay rather than being carried north to replenish the sand-starved beaches at Washaway. The tidal inlet to Willapa Bay is migrating northward, where its waves are vigorously attacking deposits of ancient sand dunes and beaches. The exact cause or causes of the erosion are uncertain. Beaches are inherently unstable and can change for small and sometimes unclear reasons. Construction of a massive jetty at the mouth of the Columbia River in 1895 interrupted sand movement that supplied the northern beaches and may have been a contributing factor. Construction of large dams on the Columbia and Snake Rivers further reduced the sediment supply by 50 percent. Rising sea level and

A trailer house hangs precariously over the retreating bluff at Washaway Beach.

other climate change factors will further influence beach processes.

Access to the beach at low tide is by the Warrenton Cannery Road. You would have found a cannery at the end of the road in the mid-twentieth century, but now you find only the crashing waves of the Pacific Ocean. The beach is strewn with concrete slabs, plumbing pipes, and other evidence of houses and buildings that have succumbed to the hungry sea. You may see buildings, trailers, and other structures hanging precariously over the 20-foot-tall bluff or on the beach below, where the unforgiving surf rips them apart. A circular concrete structure about a quarter mile out to sea from the bluff is the remnant of a World War II gun emplacement that was on the bluff edge during the early 1940s. North of the cape (accessed via Grayland Beach State Park), you'll find a beach that is growing because of the sand eroded from Washaway Beach.

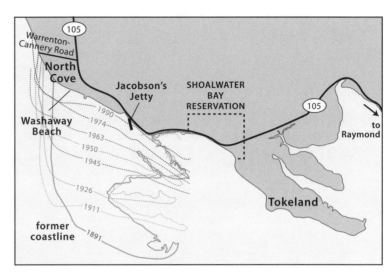

More than 2 miles of Cape Shoalwater has eroded away since the 1890s.
—Modified from Kaminsky and others, 2010

North of Cape Shoalwater, active dunes and a growing beach are benefiting from the rapidly eroding bluffs at nearby Washaway Beach.

GLOSSARY

accretion. The process of adding onto, such as adding an accreted terrane to the North American continent through plate motions.

alluvial fan. A gently sloping, fan-shaped deposit of loose rock detritus bordering the base of a steep slope and deposited by a stream or debris flow at the mouth of a canyon.

andesite. A gray to dark gray volcanic rock, made mostly of plagioclase and lesser amounts of dark minerals. Intermediate in composition and color between rhyolite and basalt.

anticline. An arch-shaped fold in layered rocks with the oldest rocks toward the core.

aquifer. A porous, permeable rock or layer of sediment that contains groundwater.

basalt. A fine-grained black or very dark gray volcanic rock consisting mainly of microscopic crystals of the minerals plagioclase feldspar and usually olivine, sometimes pyroxene.

basement. The deepest crustal rocks of a given area, typically granitic and metamorphic rocks.

batholith. A mass of granitic igneous rock exposed over an area greater than about 30 square miles and consisting of two or more plutons.

bedding. The layering structure in sedimentary rocks. A single layer is a bed. Several interlayered beds or rock types are said to be interlayered or interbedded.

bedrock. Solid rock exposed in place or that underlies unconsolidated, superficial sediments.

biotite. Black or brown mica, a flexible platy mineral. It is a minor but common mineral in igneous and metamorphic rocks.

calcite. A widespread, abundant mineral composed of calcium carbonate ($CaCO_3$). The major component of limestone and many marine fossils; a common cement, binding sediment in sedimentary rocks.

caldera. A large, steep-walled, circular or oval basin formed by collapse into an evacuated or partially evacuated magma chamber following a voluminous volcanic eruption.

carbonate rock. A sedimentary rock composed of carbonate minerals such as calcite and dolomite.

chert. A hard, dense, dull to partly glassy sedimentary rock composed mainly of microcrystalline quartz.

cirque. A steep-walled basin carved by a glacier on the side of a mountain.

clay. Rock or mineral particles smaller than 0.00016 inch; plastic when wet.

cobble. A rounded rock fragment larger than a pebble and smaller than a boulder, having a diameter in the range of 2.5 to 10 inches, or between that of a tennis ball and a volleyball.

columnar jointing. The fracturing in a lava flow that causes the flow to break into columns.

conglomerate. A coarse-grained sedimentary rock composed of rounded pebbles, cobbles, or boulders set in a fine-grained matrix of sand or silt.

continental glacier. A thick glacier that covers a large part of a continent.

coulee. In eastern Washington, a coulee is a steep-walled canyon that is dry or has a stream too small to have carved it.

crust. The uppermost, hardened layer of the Earth's lithosphere, 3 to 25 miles thick. **Continental crust** consists mainly of granitic rocks and metamorphic rocks; **oceanic crust** consists of basalt and peridotite.

dacite. A fine-grained volcanic rock with a medium-light color and a composition intermediate between andesite and rhyolite.

debris flow. A jumbled mass flow of sediment and rock fragments moving downslope.

deformation. Any process by which preexisting rocks are bent or broken.

delta. A nearly flat accumulation of clay, sand, and gravel deposited into an ocean, lake, or other standing body of water.

dike. A tabular igneous intrusion that cuts across the grain of its host rock, generally through a fracture in solid rock.

dry falls. A waterfall that no longer carries water.

epicenter. The point on the Earth's surface directly above the origin, or focus, of an earthquake.

erosion. The removal of rock material by any natural process, such as wind, water, gravity, or ice.

erratic. A rock carried by ice and deposited at some distance from where the ice picked it up.

fault. A fracture in the Earth's crust along which blocks of the crust have slipped past one another. A **normal fault** forms under crustal extension and one side drops down relative to the other side. A **thrust fault** forms under crustal shortening, and one side is pushed over the other side. A **strike-slip fault** is one where one side slips sideways relative to the other side.

feldspar. The most abundant rock-forming mineral group, making up 60 percent of the Earth's crust and consisting of calcium, sodium, or potassium with aluminum and silica.

fine-grained. A relative term used to describe the size of constituents in a rock. Said of igneous rocks with minerals too small to see with the unaided eye. Said of sedimentary rocks with silt-size or smaller particles.

flood basalt. Vast accumulations of basalt that solidified into extensive horizontal layers.

fold. Bent or warped rock layer or layers shortened by compressive tectonic stress.

fossil. The remains of a plant or animal preserved in a rock.

fracture. Any break in rocks caused by natural mechanical failure under stress, including cracks, joints, and faults.

gabbro. A dark intrusive igneous rock consisting mainly of the minerals plagioclase and pyroxene, but also olivine, in crystals large enough to be seen with a simple magnifier. Gabbro is the intrusive equivalent of basalt but contains much larger crystals because it cooled at depth over a long period of time.

glaciation. The formation and movement of glaciers or large ice sheets.

gneiss. A regionally metamorphosed rock characterized by alternating, irregular bands of coarse minerals and finer, flaky mica minerals.

graben. A crustal block that is down-dropped between two inwardly dipping normal faults.

granite. An intrusive igneous rock composed mostly of the minerals orthoclase feldspar, plagioclase, and quartz in crystals large enough to see without using a magnifier.

granodiorite. A coarse-grained intrusive rock intermediate in composition between granite and diorite. Its volcanic equivalent is dacite.

gravel. An unconsolidated natural accumulation of rounded rock fragments, most of particles larger than sand, such as cobbles and pebbles.

graywacke. A sedimentary rock made primarily of mud and sand, commonly deposited by turbidity currents.

greenschist. A schistose metamorphic rock whose green color is due to the presence of the minerals chlorite, epidote, and/or actinolite.

ice ages. Popular name for a period of Earth history, mainly during the Pleistocene epoch, when continental ice sheets periodically advanced and retreated over North America and Europe.

ice sheet. A thick glacier that covers a large part of a continent. A finger of this sheet is called an **ice lobe**.

igneous rock. Rock formed from the crystallization and solidification of molten magma.

indurated. Said of sediments consolidated or hardened by pressure, heat, and/or cement.

interglacial. A period of time between glaciations when the climate warms, vegetation returns, and the ice sheet melts back toward its source.

joint. A planar fracture or crack in a rock along which insignificant displacement has occurred.

landslide. The downslope movement of soil and rock under the force of gravity.

lava. Molten rock erupted on the surface of the Earth.

limestone. A sedimentary rock composed of the mineral calcite ($CaCO_3$).

longshore drift. Ocean current that flows near and parallel to the shore and moves sand.

magma. Molten rock within the Earth.

mantle. The largest subdivision of the Earth's interior (80 percent), which lies between the lithosphere and the very dense core.

mélange. A heterogeneous mixture of rocks.

metamorphic rock. A rock derived from preexisting rock that changes mineralogically and texturally in the solid state in response to changes in temperature and/or pressure, usually deep within the Earth. The change occurs without melting of the preexisting rock.

metamorphism. Recrystallization of an existing rock due to heat and pressure within the Earth but without the rock having melted.

microcontinent. An isolated fragment of continental crust, smaller than Australia and surrounded by oceanic crust.

Mima mound. One of many small, oval domes that occur together and form in unconsolidated sediments. Named for the Mima Prairie in western Washington.

mineral. A naturally occurring, inorganic solid substance of specific chemical composition and physical properties, or ones that vary within fixed limits.

Missoula Floods. The glacial outburst floods that swept across eastern Washington each time the ice dam holding back Glacial Lake Missoula in Montana burst.

monocline. A local steepening in an otherwise flat-lying rock layer.

moraine. A ridge of unconsolidated glacial debris deposited at the edge of a glacier as it melts. A **terminal moraine** is the moraine farthest from the source of the glacier.

mountain building. The process by which mountains rise. During this process, rocks are typically folded, faulted, and metamorphosed. Igneous activity often accompanies it.

mudflow. A mixture of water, mud, and assorted particles that flows downhill in response to gravity.

mudstone. A fine-grained sedimentary rock consisting of silt and clay.

normal fault. An inclined fault along which the overhanging rocks have moved down the inclined surface relative to the underlying rock. Forms as a result of crustal extension, or stretching.

Olympic-Wallowa Lineament (OWL). A linear topographic feature of unknown origin that stretches across Washington from the Olympic Peninsula to the Wallowa Mountains in Oregon.

outburst flood. A sudden release of water from glaciers or a glacially dammed lake.

outcrop. An exposure of bedrock at the surface. The rock is said to "crop out."

pebble. A small stone, usually worn and rounded, between 0.17 and 2.5 inches in diameter.

phyllite. A metamorphic rock that has been heated more than slate but not as much as schist; typically starts as a fine-grained sedimentary rock.

plate. In a tectonic sense, one of seven large crustal slabs composing the Earth's solid outer part. Plates are roughly 60 miles thick.

plate tectonics. A theory of global tectonics in which Earth's crust is divided into seven huge plates that move slowly and interact with one another, causing volcanic activity and earthquakes along their boundaries.

Pleistocene ice age. The last 2.6 million years of geologic time, during which periods of extensive continental glaciation alternated with warmer interglacial periods of glacial retreat.

plunge pool. A circular basin scoured by eddying currents at the base of a waterfall.

pluton. A large intrusion of magma into the solid, overlying rock. Two or more plutons make up a batholith.

pyroclastic flow. A flow of lava fragments, such as ash, cinders, blocks, and bombs, ejected during an explosive volcanic eruption.

quartz. A mineral form of silica (SiO_2). Quartz is one of the most abundant and widely distributed minerals in the Earth's crust. It comes in a wide variety of forms, including clear crystals, sand grains, and chert. It is hard and chemically resistant.

quartzite. A metamorphic rock composed of mainly quartz and formed by recrystallization of quartz-rich sandstone.

radiometric dating. The calculation of rock or sediment age from the measured ratios of one or more radiometric elements and their decay products in minerals.

resistant. Said of a rock that withstands the effects of weathering or erosion.

rift. A long, narrow belt where the Earth's crust is separating. A **rift basin** is the trough, or valley, formed by the rifting.

sand. Weathered mineral grains, most commonly quartz, between 0.0025 and 0.08 inch in diameter. Larger than silt, smaller than pebbles.

sandstone. A sedimentary rock made primarily of sand-sized particles of rock or mineral.

scabland. The dry, rocky basalt areas of eastern Washington that are lacking soil and crossed by channels scoured by the Missoula Floods.

scarp. A steep cliff face from a few feet to thousands of feet high, formed by faulting or erosion. Landslides leave behind a scarp on the mountain they slid from.

schist. A metamorphic rock with well-developed, thin layering due to an abundance of oriented, platy minerals.

sea stack. A tall outcropping of bedrock on a beach or offshore, left as a remnant of the coastline as erosion causes the coastline to retreat.

sediment. Unconsolidated rock fragments.

sedimentary rock. A rock formed from the consolidation, cementation, or compaction of loose sediment or organic material.

shale. A sedimentary deposit of clay, silt, or mud solidified into more or less solid rock that splits along parallel planes of weakness.

shearing. Deformation that occurs when two bodies of rock slide past each other, resulting in crushed rock with parallel fractures.

silica. Silicon dioxide (SiO_2); occurs naturally as quartz and as a component of many rock-forming minerals.

sill. A tabular igneous intrusion that parallels the planar structure of the enclosing rock.

siltstone. A sedimentary rock made primarily of silt- to clay-size particles.

spit. A long, narrow, fingerlike ridge of sand extending into the water from the shore.

spreading ridge. A linear ridge on the seafloor where magma wells up from below and forms new oceanic crust that, over time, spreads away from the ridge.

strike-slip fault. A fault along which the relative displacement is sideways rather than up or down.

subduction zone. A type of convergent plate tectonic boundary where oceanic crust of one plate descends beneath another plate.

syncline. A downfold in layered rocks with younger beds toward the core, limbs inclined inward.

tectonic. Pertaining to the deformation of the Earth's surface, including plate movements, due to internal Earth forces.

terrace. Steplike landform consisting of a flat tread and a steep riser, commonly shaped by the running water of rivers or the waves along the shores of lakes and oceans, and standing above the present river or coast.

terrane. Fault-bounded crustal fragment with a geologic history that differs from adjacent fragments.

thrust fault. A fault that dips less than 45 degrees, upon which the older, upper block of rocks moves up and above the younger rocks beneath the fault. Typically forms in crustal shortening. The block of land that is moved is called a **thrust sheet**.

till, glacial. An unlayered and unsorted mixture of clay, silt, sand, gravel, and boulders deposited directly by a glacier.

tonalite. An intrusive igneous rock composed mostly of plagioclase feldspar and from 5 to 20 percent quartz. Tonalite grades into granodiorite as the alkali feldspar content increases.

tuff. A volcanic rock made mostly of consolidated pyroclastic volcanic ash and pumice.

turbidity current. A dense slurry of sediment that moves down an underwater slope.

unconformity. A surface of erosion and/or nondeposition separating younger deposits from underlying older rocks, and across which the rock record is missing.

volcanic arc. An arcuate chain of volcanoes that formed above a subduction zone.

volcanic rock. Extrusive, fine-grained igneous rock, cooled and solidified after eruption onto the surface of the Earth.

water gap. A deep gorge or ravine cut through resistant rocks by a stream.

weathering. The mechanical breakdown or chemical decomposition of rocks and minerals at the Earth's surface in response to atmospheric agents such as air and rain.

welded tuff. A glassy pyroclastic volcanic rock that has been hardened by the fusing together of its glass shards under the combined action of heat, weight of overlying rocks, and hot gases.

REFERENCES AND FURTHER READING

Allen, J. E. 1979. *The Magnificent Gateway*. Timber Press.

Allen, J. E., M. Burns, and S. Burns. 2009. *Cataclysms on the Columbia*. Ooligan Press.

Alt, D. 2001. *Glacial Lake Missoula and Its Humongous Floods*. Mountain Press.

Alt, D., and D. W. Hyndman. 1984. *Roadside Geology of Washington*. Mountain Press.

Alt, D., and D. W. Hyndman. 1995. *Northwest Exposures: A Geologic Story of the Northwest*. Mountain Press.

Atwater, B., M. R. Satoke, T. Yoshinobu, U. Kazue, and D. Yamaguchi. 2005. *The Orphan Tsunami of 1700*. US Geological Survey Professional Paper 1707.

Ayden, A., and J. M. DeGraff. 1988. Evolution of polygonal fracture patterns in lava flows. *Science* 239: 471–76.

Babcock, S., and B. Carson. 2000. *Hiking Washington's Geology*. The Mountaineers.

Berg, A. W. 1990. Formation of Mima mounds: A seismic hypothesis. *Geology* 18(3): 281–84.

Bjornstad, B. 2006. *On the Trail of the Ice Age Floods: Mid-Columbia Basin*. Keokee Publishing.

Bjornstad, B., and E. Kiver. 2012. *On the Trail of the Ice Age Floods: The Northern Reaches*. Keokee Publishing.

Blakely, R. J., B. L. Sherrod, C. S. Weaver, R. E. Wells, and A. C. Rohay. 2014. The Wallula fault and tectonic framework of south-central Washington, as interpreted from magnetic and gravity anomalies. *Tectonophysics* 624: 32–45.

Brown, N. 2014. *Geology of the San Juan Islands*. Chuckanut Editions.

Carson, B., and S. Babcock. 2009. *Hiking Guide to Washington Geology*. Keokee Publishing.

Conner, C. 2014. *Roadside Geology of Alaska*, second edition. Mountain Press.

Daniels, R. C., R. H. Huxford, and D. McCandless. 1998. *Coastline Mapping and Identification of Erosion Hazard Areas in Pacific County, Washington*. Department of Ecology, Coastal Monitoring and Analysis Program, Olympia, Washington.

DeCelles, P. G. 2004. Late Jurassic to Eocene evolution of the Cordilleran thrust belt and foreland basin system, western USA. *American Journal of Science* 304: 105–68.

Dotson, C. L., ed. 2010. *Pacific Northwest Ice Age Floods Field Trip Guides: Eastern Washington Landscape*. Ice Age Floods Institute.

Doughty, T. P., and R. A. Price. 1999. Tectonic evolution of the Priest River complex, northern Idaho and Washington: A reappraisal of the Newport fault with new insights on metamorphic core complex formation. *Tectonics* 18: 375–93.

Fuentes, F., P. G. DeCelles, K. N. Constenius, and G. E. Gehrels. 2011. Evolution of the Cordilleran foreland basin system in northwestern Montana, USA. *Geological Society of America Bulletin* 3: 507–33.

Galm, J. R., and K. McClure-Mentzer, eds. 2009. *Long Mound: A Late Holocene Archaeological Site in the Channeled Scablands of Eastern Washington*. Report Series No. 8, Department of Geography and Anthropology, Eastern Washington University.

Greely, R., and J. H. Hyde. 1972. Lava tubes of the Cave Basalt, Mount St. Helens, Washington. *Geological Society of America Bulletin* 83(8): 2397–2418.

Harris, S. L. 1980. *Fire and Ice*, revised edition. The Mountaineers and Pacific Search Press.

Ice Age Floods Institute. 2010. *Ice Age Floods in the Pacific Northwest*. Keokee Publishing.

Johnson, S. Y., C. J. Potter, J. J. Miller, J. M. Armentrout, C. Finn and C. S. Weaver. 1996. The Southern Whidbey Island fault: An active structure in the Puget Lowland, Washington. *Geological Society of America Bulletin* 108: 334–54.

Kaminsky, G. M., P. Ruggiero, M. Buijsman, D. McCandless, and G. Gelfenbaum. 2010. Historical evolution of the Columbia River littoral cell. *Marine Geology* 273: 96–126.

Kiver, E. P., and D. V. Harris. 1999. *Geology of US Parklands*. John Wiley and Sons Publishing.

Lipman, P. W., and D. R. Mullineaux, eds. 1981. *The 1980 Eruptions of Mount St. Helens, Washington*. US Geological Survey Professional Paper 1250.

Martin, K. 1988. Oxygen isotope analysis, uranium-series dating, and paleomagnetism of speleothems in Gardner Cave, Pend Oreille County, Washington. Thesis. Eastern Washington University.

Mathews, B., and J. Monger. 2010. *Roadside Geology of Southern British Columbia*. Mountain Press.

Miller, M. B. 2014. *Roadside Geology of Oregon*, second edition. Mountain Press.

Mustoe, G. E. 2010. Biogenic origin of coastal honeycomb weathering. *Earth Surface Processes and Landforms* 35(4): 424–34.

Pogue, K. R. 2009. Folds, floods, and fine wine: Geologic influences on the terroir of the Columbia Basin. In *Volcanoes to Vineyards: Geologic Field Trips through the Dynamic Landscape of the Pacific Northwest*, J. E. O'Connor, R. J. Dorsey, and I. P. Madin, eds. Geological Society of America Field Guide 15, 1–17.

Pringle, P. T. 1993. *Roadside Geology of the Mount St. Helens National Volcanic Monument and Vicinity*. Washington Division of Geology and Earth Resources Information Circular 88.

Pringle, P. T. 2008. *Roadside Geology of Mount Rainer National Park and Vicinity*. Washington Division of Geology and Earth Resources Information Circular 107.

Rau, W. W. 1980. *Washington Coastal Geology*. Washington Department of Natural Resources.

Reidel, S. P. 2006. *Big Black Boring Rock*. Battelle Press.

Reidel, S. P., V. E. Camp, T. L. Tolan, and B. S. Martin. 2013. The Columbia River flood basalt province: Stratigraphy, areal extent, volume, and physical volcanology. In *Columbia River Flood Basalt Province*, S. P. Reidel, V. E. Camp, M. E. Ross, J. A. Wolff, B. S. Martin, T. L. Tolan, and R. E. Wells, eds. Geological Society of America Special Paper 497.

Sissons, T. W., and M. A. Lanphere. 1999. The geologic history of Mount Rainier volcano, Washington (abstract). In *A Century of Resource Stewardship and Beyond: Mount Rainier National Park 100th Anniversary Symposium*. Northwest Scientific Ass.

Sissons, T. W., and J. W. Vallance. 2009. Frequent eruptions of Mount Rainier over the last 2,600 years. *Bulletin of Volcanology* 71: 595–618.

Slotemaker, T. 2013. *The Geology of Fidalgo Island*. Anacortes Museum.

Soennichsen, J. 2008. *Bretz's Flood*. Sasquatch Books.

Tabor, R. W. 1975. *Guide to the Geology of Olympic National Park*. University of Washington Press.

Tabor, R., and R. Haugerud. 1999. *Geology of the North Cascades*. The Mountaineers.

Tucker, D. 2015. *Geology Underfoot in Western Washington*. Mountain Press.

Veatch, F. M. 1969. *Analysis of a 24-Year Photographic Record of Nisqually Glacier, Mount Rainier National Park, Washington*. US Geological Survey Professional Paper 631

Washburn, A. L. 1988. *Mima Mounds: An Evaluation of Proposed Origins with Special Reference to the Puget Lowland*. Washington Division of Geology and Earth Resources Report of Investigations 29.

Watkinson, A. J., and M. A. Ellis. 1987. Recent structural analyses of the Kootenay Arc in northeastern Washington. *Washington Division of Geology and Earth Resources Bulletin* 77: 1–14.

Williams, H. 2002. *The Restless Northwest*. Washington State University Press.

Winston, D., and P. K. Link. 1993. Middle Proterozoic rocks of Montana, Idaho and eastern Washington: The Belt Supergroup. In *Precambrian of the Conterminous United States*, J. C. Reed et al., eds., Geological Society of America, Geology of North America C-2: 487–517.

Websites

Washington Department of Natural Resources Geologic Information Portal. http://www.dnr.wa.gov/geology.

INDEX

Eugene Kiver, Richard Orndorff, and Chad Pritchard

Eugene Kiver, an emeritus professor of geology who taught and did research on geomorphology and ice age history at Eastern Washington University for thirty-two years, now lives in the seaport of Anacortes, Washington. He is on the Ice Age Floods Institute Board, leads field trips in the Channeled Scabland, and makes numerous presentations to interested groups. He recently coauthored *On the Trail of the Ice Age Floods* with Bruce Bjornstad. Gene is an active hiker and has hiked well over 3,000 miles in the past ten years in the Cascade Range and elsewhere in the Pacific Northwest.

Chad Pritchard, a geology professor at Eastern Washington University, grew up rock climbing, snowboarding, and Scuba diving throughout the Northwest. After various jobs in the earth sciences, including with the soil survey in Spokane and as an environmental regulator in Hawaii, he obtained a PhD in geology from Washington State University. In addition to studying a great variety of geologic features in Washington, Chad has researched geothermal heat in Iceland, volcanics in Yellowstone, and land level changes associated with large Cascadia earthquakes on the West Coast. He lives with his wife and three children in Medical Lake, Washington.

Richard L. Orndorff, professor of geotechnical engineering at Eastern Washington University, is a coauthor of *Geology Underfoot in Central Nevada, Geology Underfoot in Southern Utah,* and *Landforms of Southern Utah,* all published by Mountain Press Publishing Company. He resides in Spokane with his wife, Sherry, daughter, Emma, and dog, Chloe.